Value-Based Management

Delivering Superior Shareholder Value

Value-Based Management

*Delivering Superior
Shareholder Value*

**GARY ASHWORTH
PAUL JAMES**

An imprint of Pearson Education

London ■ New York ■ San Francisco ■ Toronto ■ Sydney ■ Tokyo ■ Singapore
Hong Kong ■ Cape Town ■ Madrid ■ Paris ■ Milan ■ Munich ■ Amsterdam

PEARSON EDUCATION LIMITED

Head Office:
Edinburgh Gate
Harlow CM20 2JE
Tel: +44 (0)1279 623623
Fax: +44 (0)1279 431059

London Office:
128 Long Acre
London WC2E 9AN
Tel: +44 (0)20 7447 2000
Fax: +44 (0)20 7240 5771
Website: www.briefingzone.com

First published in Great Britain in 2001

ISBN 0 273 65404 7

British Library Cataloguing in Publication Data
A CIP catalogue record for this book can be obtained from the British Library.

This book contains a family of share valuation metrics known as Enterprise
Value Added, EV+. These are patented by Sharevaluer and may only be used
under licence obtainable through Sharevaluer.com.

10 9 8 7 6 5 4 3 2 1

Typeset by Monolith – www.monolith.uk.com
Printed and bound in Great Britain

The Publishers' policy is to use paper manufactured from sustainable forests.

About the authors

 **Gary Ashworth** is the founder and managing director of Knowledge-Capita, a company specialising in the provision of training, coaching and consultancy on advanced business performance management techniques. He has a particular interest in ensuring that the tools and techniques he implements have real and lasting business impact. He has extensive implementation experience in some of the UK and Europe's leading companies and he is regularly invited to speak at conferences and seminars.

A former leader of Ernst & Young's Performance Management Service Line in the UK for some five years and the previous holder of a similar role on Activity Based Management at KPMG, he has specialised in advanced business performance management approaches as a management consultant for over ten years. While at Ernst & Young, Gary directed the development and delivery of VBM training in Europe.

His passion for improving approaches to managing business performance is born out of both his commitment to his profession and his wide range of practical experience prior to entering consultancy. This included working for over 12 years at senior management levels in companies such as Shell, Whitbread and Sainsbury's.

He has written, and had published, many articles on performance management over recent years. A good number of these have been given international acclaim by the International Federation of Accountants (IFAC) for making 'a distinguished contribution' to the development of the management accounting profession. Indeed, one article won IFAC's 'First Place' award in the worldwide published article competition of that year. This is his second publication in the Financial Times Management Briefings series.

He is a Fellow of the Chartered Institute of Management Accountants (and twice an Institute Prize winner), a Fellow of the Institute of Management Consultants, a Certified Management Consultant and Member of the British Institute of Management.

He can be contacted at GA@knowledge-capita.com or by phone on +44 (0)118-925 3377.

 **Paul James** is a specialist in share valuations and the data used by analysts in the investment community. Through his company, Enterprise Value Management Consultants, he works for many of the leading providers of investment information and tools. His clients include Thomson Financial Datastream, First Call-I/B/E/S, royal**blue**, Thomson Financial PGSE and Reuters.

During his career, Paul has held a number of senior financial positions in industry, working for companies such as ABB, P&O and Tarmac. Paul has also spent 11 years in management consultancy for PWC, Cap Gemini/Ernst & Young and BDO, where he was partner in charge of financial management consulting.

Paul is the founder and managing director of Sharevaluer. Sharevaluer focuses on the professional investment community and its mission is to deliver the competitive edge in investment information to professional fund managers and analysts.

He has invented and patented an advanced residual income share valuation technique, Enterprise Value Added (EV+). This family of measures is designed to provide professional investors with new insights into the pricing of shares and sectors. These valuation metrics are marketed through Sharevaluer.

Paul is frequently asked to speak at public conferences. His focus is normally on the external investment community and its perception of shareholder value. During these presentations he also explains the application of EV+ valuation methods to different sectors and stocks.

Paul has a degree in economics, is a Fellow of the Institute of Chartered Accountants and a Member of the Institute of Management Consultants.

He can be contacted by e-mail at pjames@sharevaluer.com or by phone on +44 (0)7967 645978.

Contents

List of figures

Figures 5.1–5.3, 5.5–5.7, 5.9, 5.10, 6.1–6.5, 7.1–7.3 have been adapted from Gary Ashworth's *Delivering Shareholder Value Through Integrated Performance Management* (1999) Financial Times Publications.

Original figures for Figure 5.1, 5.9, 6.1–6.5 © Ernst & Young.

List of tables

Preface

The creation and delivery of shareholder value has become a business mantra espoused by almost every self-respecting chief executive. In their annual reports and published results few of them fail to mention their focus on 'delivering shareholder value'. However, many organisations fail to translate the aim into reality. Some manage to develop a strategy for creating value. Few actually deliver.

In the increasingly e-connected economy, investors move their money quickly and easily around the world in the quest for the optimum shareholder returns. As a result, today's business leaders and managers must be able to understand how to create, measure, manage and deliver shareholder value. Messages about value in annual reports are not enough on their own.

Too often, the focus is on a few key measures that in themselves deliver little real or lasting business benefit. Such companies fail to present a clear and consistent view of what shareholder value means or why it is important. They fail to create an approach that is consistent, comprehensive and quality assured. They also miss the point that value, and the way in which it will be delivered, needs to be communicated both internally and externally.

Addressing these issues successfully involves the adoption of a true value-based management approach within a strong value culture. For those who have done this, the results have been both impressive and consistent over time.

The key to delivering a vision for value is to implement an integrated approach that aligns internal business activities with external shareholder expectations. This book gives some unique insights into how this can be achieved by focusing on three interlinked components of value-based management. These components not only help companies with their strategic thinking, but also with external and internal communication on how value will be created. The book then moves on to describe how to translate strategic vision into commercial reality.

Acknowledgements

We would like to thank the following:

- Alan Warner, Founder and Chairman of Management Training Partnership. Alan made a significant contribution to the text of this book. He also drew upon his clients' experiences in implementing VBM which has added spice to some of the VBM concepts described herein.

- The companies to whom we have felt able to refer, in particular BP and Peter Hall's contribution. These also include Cadbury Schweppes, Unilever, Boots, Lloyds TSB, BA, Cargill, Invensys, Sony, Barclays, and Sainsbury's.

- Thomson Financial for the input made to the text by their investor relations consultants and for the charts they provided through Thomson Financial Datastream.

- ING Barings and Paribas for permission to use their broader research in Appendix 5.

We also wish to acknowledge the trademark ownership of certain terms used in the book. These include:

- EVA, Economic Value Added, and MVA, Market Value Added, which are trademarks of Stern Stewart;

- CFROI, Cash Flow Return on Investment, which is a trademark of Holt Value Associates, LP;

- EV+, Enterprise Value Added, which is patented set of valuation measures from Sharevaluer.

List of abbreviations

ABC	activity-based costing
CAPM	capital asset pricing model
CEO	chief executive officer
CFO	chief financial officer
CFROI	cash flow return on investment
CPS	cash flow per share
CSF	critical success factor
DDM	dividend discount model
DCF	discounted cash flow
DPS	dividends per share
EBITDA	earnings before interest, tax, depreciation and amortisation
EIS	executive/enterprise information system
EPS	earnings per share
EV	enterprise value
EV+	enterprise value added
EVA	economic value added
FCF	free cash flow
FTSE	Financial Times Stock Exchange Index
IPM	integrated performance management
IR	investor relations
IRR	internal rate of return
KPI	key performance indicators
MIS	management information system
MVA	market value added
NOPAT	net operating profit after tax
NPV	net present value
PBT	profit before tax
P/E (PER)	price/earnings ratio
PEG	price earnings growth
PV	present value
R&D	research and development

ROCE   return on capital employed

SEC   Securities and Exchange Commission

SME   small or medium-sized enterprise

TBR   total business return

TSR   total shareholder return

VBM   value-based management

WACC   weighted average cost of capital

To Mandy, Marcelle, Lauren, Dominic, Sean, Christopher and Andrew.

Shareholder value and value-based management

CHAPTER SUMMARY

- Creating and delivering shareholder value demands that shareholders get a return over the cost of their capital through dividends and increased business value.

- A number of major companies have adopted a management technique called value-based management (VBM) to help create superior shareholder value.

- Several of these leading VBM exponents have consistently delivered superior levels of shareholder value.

- As a result, several other well-known companies are also now looking to emulate this success through a focus on VBM. This suggests that it is now recognised as a proven management technique for delivering improved business performance.

- At its most basic level, a successful VBM approach means achieving a long-term return on capital in excess of its cost. This can be easily calculated as an internal financial metric and is often referred to as economic profit.

- However, achieving strong and sustainable shareholder value through VBM requires more than just the calculation of one internal financial metric. There are three key distinct but inextricably linked perspectives or processes of managing for shareholder value or adopting a VBM approach. These are defined as:
    - managing the investor community;
    - evaluating strategies to create optimal value;
    - delivering value through integrated performance management.

- These three perspectives or integral executive management processes form the overarching structure around which the focus for VBM in this book is built. In other words, the central theme is that VBM is all about harnessing these three management processes in harmony at the top executive level to maximise and deliver superior shareholder value.

- This is achieved, in part, by using an appropriate choice of value-based metric(s) linked to other financial and non-financial measures. Together these form a common language of 'value' and they help to give clear direction on what has to be done throughout the organisation when embedded successfully in the business.

If you are going to strive to be number one in shareholder value then you have got to be better than all the rest. The shareholder value concept is embedded within our business, not just spread over the top. It comes down to the organisation structure we have – and everyone buys into their own piece, to the targets they've agreed and signed off.

Peter Hall, Director of Investor Relations, UK & Europe – BP

INTRODUCTION TO THE BUSINESS RATIONALE OF VBM

Finance experts argue that companies need to earn a minimum level of return on all the capital they employ within their organisations. This minimum level of return required by the providers of capital is known as the 'cost of capital'. This means that after paying the providers of debt capital, there must still be enough left in order to compensate the equity shareholders for the risks that they take.

The returns to shareholders can take the form of dividends and growth in the value of their shares. However, in the long run, unless companies are able to deliver returns that exceed the cost of capital, then the shareholders will grow dissatisfied, disposing of their investments and forcing down the share price.

Falling share prices erode the value of equity investments and lead to disgruntled investors. Disgruntled investors, if upset for long enough, may seek to replace existing managers with those who can produce results of the size needed to maintain and increase share price. There is strong evidence[1] of increasing shareholder activism of this sort. Understandably companies, and their executive management teams, seek tools to help them measure and deliver value to shareholders.

VBM is such a management technique. It is designed to help companies create superior shareholder value through aligning the focus of management decision-making with the interests of shareholders. The success of the technique appears to be leading to an increasing trend towards its adoption. For example, in recent times, it has become public knowledge that major companies like Barclays and Sainsbury have started to focus on VBM to help them manage and, indeed, transform their businesses.

Can there be any doubt as to why the popularity of VBM is increasing? If there is, then one only has to look at the relative performance of these companies in delivering shareholder value to determine a significant part of the answer. Figures 1.1 to 1.3 below show several sets of comparisons to illustrate the point. All are rebased to a common start to aid the comparison.

Figure 1.1 shows the share price performance of Lloyds prior to the time of its merger with TSB in the mid-1990s. Lloyds first came to adopt a VBM approach in the mid-1980s with the appointment of Sir Brian Pitman. The graph compares the share price of Lloyds with a major competitor, Barclays, and also with the Datastream Banks Index over a 15-year period. The top line of the graph represents Lloyds and shows clearly how impressively it performed in relation to its peers. Given the relative performance of Barclays over time, it is no wonder that Barclays announced the introduction of VBM in 2000 with the express aim of helping it to become a top-tier performer.

However, the effect of a focus on managing for shareholder value is not just being felt within the banking sector. It applies equally in a variety of different industries. The graph in Figure 1.2 illustrates its impact on the oil and gas sector.

Fig. 1.1 The Lloyds share price trend

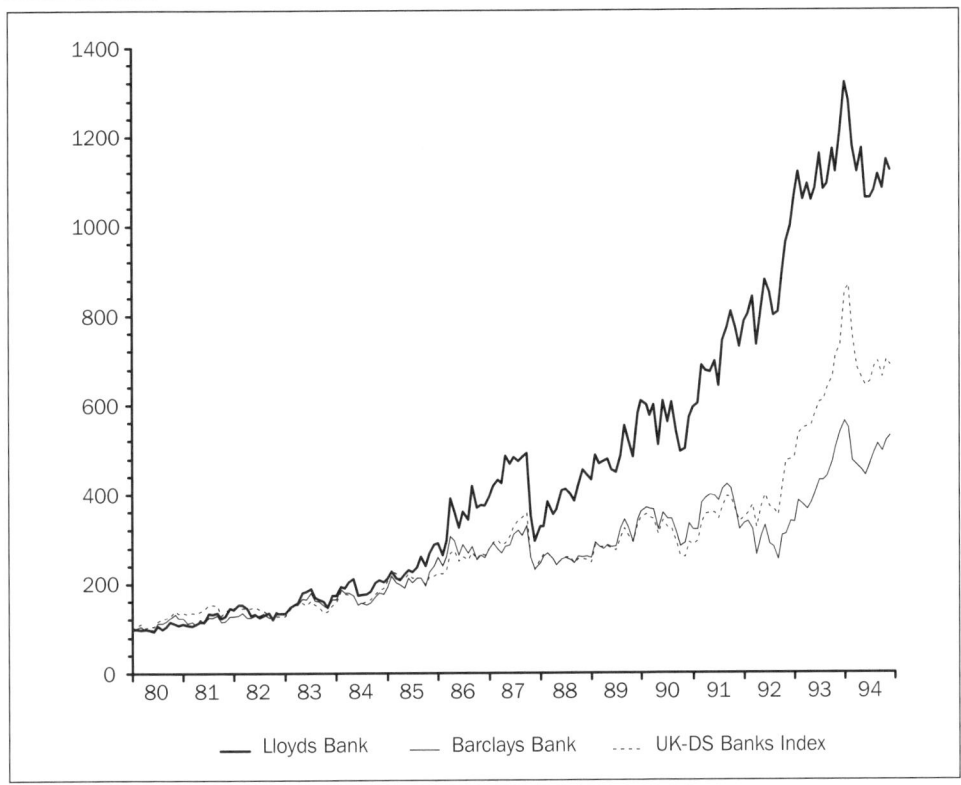

Source: Thomson Financial Datastream

Fig. 1.2 The BP share price trend

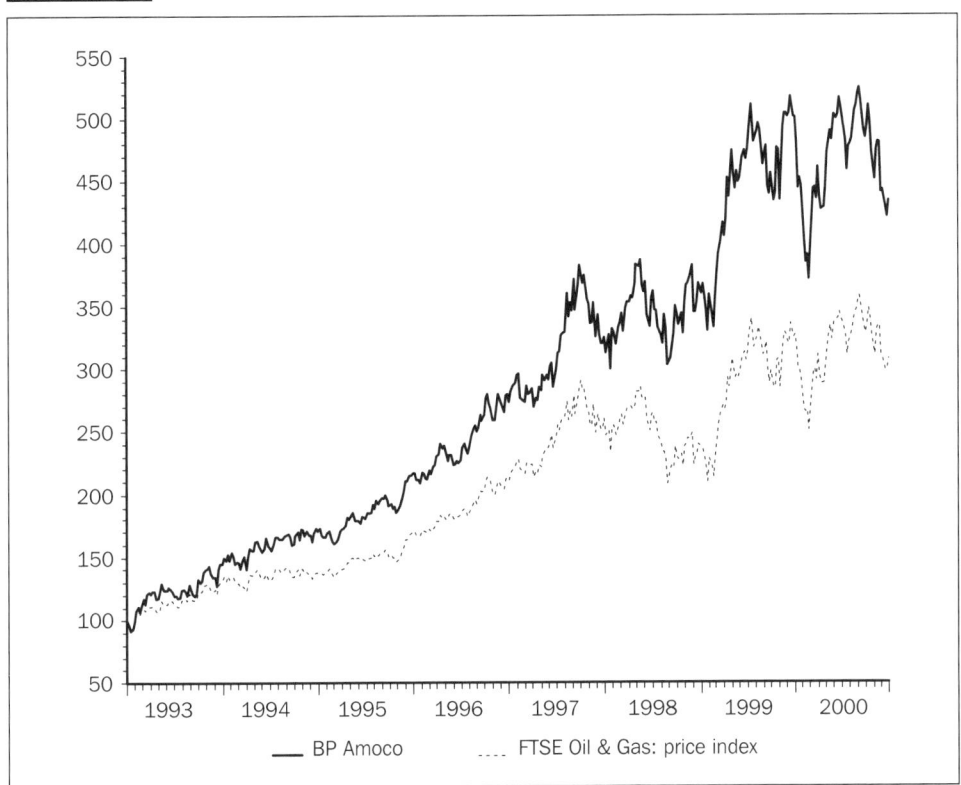

Source: Thomson Financial Datastream

Figure 1.2 shows a relative share price graph for BP. Since the early/mid-1990s BP has focused its efforts on creating a total return to shareholders that is in excess of its peers. This graph shows the comparison of BP's share price growth with the FTSE Oil and Gas index for that period. The increasing difference between the two lines demonstrates BP's success in maintaining the delivery of superior shareholder value. BP's approach is described in the case study later in this chapter.

The retail sector has also seen companies that manage for shareholder value. Figure 1.3 illustrates the relative share price performance of two UK household names in the retailing of consumer products, Boots and Sainsbury. Boots began adopting VBM principles in the early 1990s and since that time their shares have delivered superior shareholder value.

Fig. 1.3 The Boots share price trend

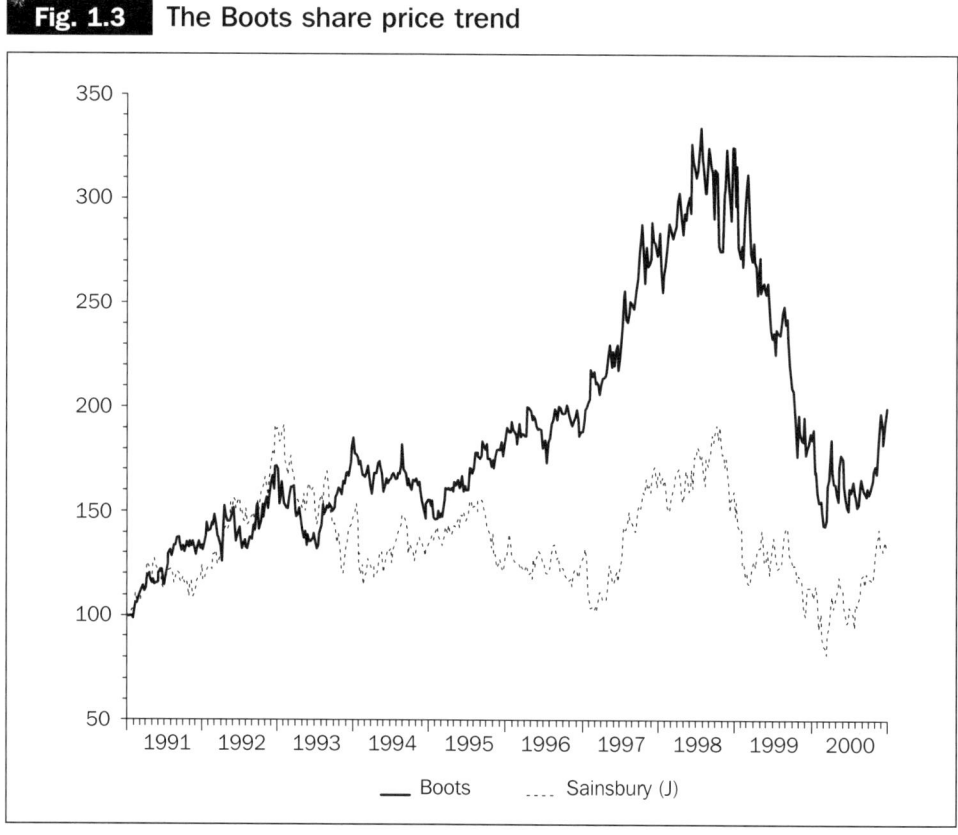

Source: Thomson Financial Datastream

VBM alone cannot be proclaimed as the single source of all this superior performance, but it is generally accepted as a key ingredient. This is because it generates a greater clarity of purpose in the companies that adopt it. While VBM has been implemented differently in these companies, it is most certainly a contributory factor in helping them to deliver superior shareholder value. The shareholder value concept is thoroughly embedded in each of these organisations.

It is not surprising therefore that Barclays and Sainsbury are now looking to transform their performance in the same way their competitors have done. They are aiming to do this, at least in part, through a focus on this management tool we call VBM.

These comparisons are but a small sample. However, they serve to highlight the benefits being gained by leading British exponents of VBM. Further, of the top ten 'most admired' companies in Britain, at least five are believed to have adopted a value-based approach and are realising benefits from it. This is in keeping with further evidence of VBM success in the US, with Coca-Cola its most famous exponent on the global stage.

Shareholder value became a business mantra in the 1990s and yet it is likely to become even more widely espoused in the new millennium. Why? The increasing focus on delivering shareholder value is clear. The focus on value creation gives a sense of purpose for energising high performance in every aspect of the business.

The old saying 'what gets measured gets done' is certainly true in the world of shareholder value. When businesses manage for shareholder value they tend to adopt a common language of value-based metrics. These can in turn be linked to other financial and non-financial measures and targets which help to drive success and, importantly, deliver superior returns for investors when embedded successfully in the business.

MEASURING VALUE CREATION

The metrics that measure 'value' or 'value creation' were originally based on discounted cash flow (DCF) techniques and these are most commonly applied to individual project evaluations. For completeness a simple explanation of the principles of DCF is included as Appendix 1.

The first step in measuring the value created from any investment project is to calculate the net present value (NPV). The NPV represents:

- The sum of the 'present values' of future cash flows resulting from an investment that is discounted at a given rate of interest, the 'cost of capital'. (This gives a sum for the future receipts from the investment, expressed in today's monetary values.)

- Less the cost of the investment. (This determines whether, again in today's money, there is a surplus or deficit from the investment.)

The NPV that results, therefore, simply represents the 'present value' of the future cash flows less the original cost of the investment. If the NPV is positive then the return from the investment has exceeded the cost of capital and *the value of the company should increase by the amount of the value created.* If, however, the NPV is negative the company's value should theoretically decrease.

In reality there are many complications to this simple scenario. Companies represent a composite portfolio of numerous investment projects that have been made at different points in time and they do not convey all the information investors need to adjust values accordingly. Further, complex investment project scenarios can be extremely difficult to analyse and there are many arguments about the correct discount rates to use.

In spite of the difficulties, and although investors cannot always delve into the results of individual projects, it is possible for them to study the accounts of companies and to infer from them whether value has been added or destroyed. Investors can also extend this approach further by analysing the forecasts for companies to determine whether they are likely to add value in the future. This can help with their investment decisions and will, in turn, affect share prices.

Strictly speaking, therefore, companies wishing to deliver and maximise shareholder value creation only need to focus on two things:

- maximising the stream of future cash flows;
- minimising the interest charged against that stream by reducing the 'cost of capital'.

In fact, some would argue that influencing the cost of capital charge significantly is almost impossible and that the sole focus should therefore be on cash flow maximisation. At its most basic level then, a successful VBM approach means achieving a positive stream of future cash flows to give shareholders a return on capital in excess of its cost.

There are many ways of measuring this value but two lead the rest of the pack, being the most well known. The first is *total shareholder return* and the second is *economic profit*.

TOTAL SHAREHOLDER RETURN AND ECONOMIC PROFIT

From an investor's perspective, when measuring the value that has been created, the most important measure to use is total shareholder return (TSR). TSR is the sum of two components which represent the benefits to the shareholder from owning the share:

- the percentage share price appreciation over the period being measured;
- the dividend yield realised during the period, again expressed as a percentage of the share price.

Internally, VBM companies often adopt the second most popular measure of value. This is known as 'economic profit'. It measures the return (or profit) earned by the company in a period after deducting a charge for the cost of the capital employed within the business. Economic profit is often considered as the internal VBM measure which acts as a proxy, in the long run, for the shareholder value measured externally by TSR.

These two terms are often most closely associated with VBM, hence their introduction here at the outset. However, these terms represent a narrow view of value creation and the measurement of shareholder value. To deliver superior shareholder value consistently companies need to consider a broader perspective. For example, there are many other important measures of value. These are explored in Chapter 2 which is devoted entirely to this topic.

Companies that have successfully embraced the concept of VBM and delivered superior shareholder value are worthy of analysis. This is because they can teach us some of the techniques they have applied and provide a role model for future success.

A CASE STUDY IN DELIVERING
SHAREHOLDER VALUE – BP

BP is one of the few companies that regularly receive awards for its delivery of shareholder value. Peter Hall is often the recipient of these awards. He is the Director of Investor Relations at BP with specific responsibility for the UK and Europe.

Peter highlights a number of key factors that keep BP near the top of the shareholder value league tables:

- The concept of shareholder value is very important within the business culture. The group actively attempts to manage and integrate the shareholder value perceived externally within the stock markets with the value created internally by the managers of the business. There is a very close link between the investor relations team and the Group CEO and CFO, who continually take a strong interest in the company's share price. The group's investor relations department is eight strong and has dedicated experts both in London and New York.

- The company has adopted TSR as its main way of measuring of shareholder value. Absolute growth in TSR is not deemed to be sufficient. BP must also improve relatively against its peer group. BP has also used TSR in a sophisticated way by:
  - calculating TSR return over a three-year period to smooth out short-term fluctuations in the stock markets;
  - using a comparative peer group of competitors (against whom they measure their relative TSR performance). This group includes the major seven key players within the oil industry – Exxon/Mobil, Shell, Texaco, Chevron, Total/Fina/Elf, ENI and Repsol YPF.

- TSR has been adopted internally as a way of driving business performance. This has been achieved though the use of two additional internal performance measures, earnings growth coupled with return on capital employed. The concept is that the two measures together form a useful internal equivalent of TSR. Peter argues that, by setting business unit targets based on both of these measures, managers need to deliver growth in earnings and an increasingly effective utilisation of the assets within their businesses. Success with these parameters should translate into improved TSR.

- The organisation structure is deliberately flat and is geared towards effective management and delivery against these key targets. There are approximately 150 business units in BP, each on average with around $0.5 billion in capital employed. This is small enough to enable the CEO, if he wants to communicate an important message, to bring together all the managers of the business units into one room if necessary.

- The business units within BP also compete against each other for capital allocation. Peer group results are regularly reported through BP's financial systems. This internal competition also drives selective investment into those business projects that generate the best shareholder value in the medium to long term.

- The remuneration of the entire management team is heavily dependent upon their relative performance against demanding three-year TSR, earnings growth and return on capital employed (ROCE) targets. It is not enough for managers to deliver TSR, earnings growth or ROCE in isolation. To earn the maximum award they have to deliver all three and they have to outperform their relevant peer group. Performance below the peer group median results in no award. The remuneration contract of BP's executive directors tie as much as 70% of their earnings to these factors so they really do bear similar risks to the shareholders for whom they act. The contracts of all other senior managers are similarly structured to reflect both their own targets and the company's results relative to its peers.

Peter points out how essential it is for him to remain close to BP's businesses in order to do his job effectively. The investor relations team maintains a huge network of close contacts within each of the businesses. He takes real pride in this and argues that, in order to communicate properly to analysts and investors, he needs to be able to understand and explain the economic realities behind BP's business performance.

BP also recognises the need to manage effectively its communications with journalists and the investment community. It is important that the releases generated by BP's Press Office complement, and do not conflict with, the messages the investor relations team are giving.

Peter's team attempts to actively monitor the information that is currently available about BP within the investment community. BP subscribes to Reuters, Bloomberg, Thomson Financial Datastream and First Call-I/B/E/S. This enables Peter's staff to keep under review the data held within the systems of these major information providers. From these data sources BP also maintains an up-to-date picture of the analysts' consensus forecasts for the company.

Often analysts will send drafts of their reports and models to the company for comment. BP is frequently written about so the number of these reports can be significant. The team attempts to review these to correct any factual errors but they do not comment on the brokers' recommendations. This can provide an opportunity to clarify certain details, including those on strategy, where the analyst may not fully have grasped all the issues.

BP is also in regular contact with large institutional shareholders, such as Fidelity and Merrill Lynch Asset Management, who by virtue of the amount of their funds under management can make a significant difference to BP's share price performance relative to its competitors within the industry.

This BP case study highlights an important truth about managing for shareholder value. It is not sufficient for companies merely to convey to investors messages about shareholder value. They must also back up those messages with a real implementation of practical measures and actions designed to create and deliver that value consistently.

Understanding the relationship between internal and external perspectives on value is therefore a key challenge in adopting a successful VBM approach.

THE THREE KEY PERSPECTIVES OF VBM

Too often VBM is narrowly construed as a focus on one or two financial metrics, such as economic profit or total shareholder return, and their related targets. However, while some measures are better suited for managing or understanding investors' expectations, others can help evaluate different strategies and measure the potential value they may create, and yet others can help measure internal management performance and may be more appropriately linked to reward and compensation schemes.

Companies therefore need to take a 'fit for purpose' approach, reflecting their own particular circumstances and the way they wish to apply a particular measure to their own business management processes. No single metric is right for all purposes and, whatever approach is taken, the key is to obtain clarity about the measures used and their linkage to the creation of shareholder value.

Measures alone are not enough to deliver value from a VBM initiative. Yet well-defined metrics can help to measure the creation of value and encourage managers to focus on its delivery. They are also a means of exploring how best to create value and communicate the delivery of superior shareholder value both internally and externally. This is because the metrics provide the common language that is needed to address issues within three distinct but strongly linked areas. These are highlighted in Figure 1.4.

Fig. 1.4 The three perspectives of VBM

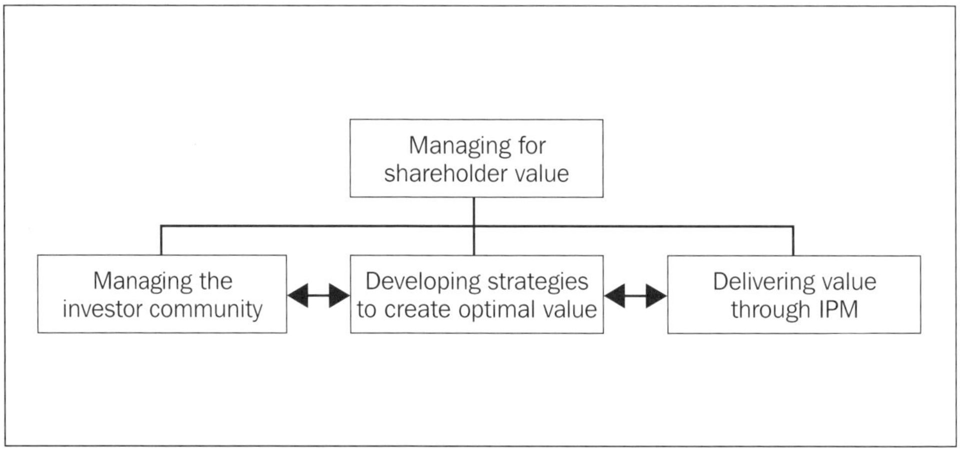

Managing the investor community

This management process demands that companies maintain effective relationships and good communications with the investor community in order to support their share price. It requires a thorough understanding of both the financial and non-financial performance factors that matter to investors. Financial performance is obviously very important but empirical evidence shows that a number of non-financial factors are also taken into account when investors value companies. The market's perception of these factors can have a very significant impact on value and how well they are managed, such as, for example, the perception of management's ability to execute strategy.[2]

Evaluating strategies to create optimal value

This management process requires a systematic analysis aimed at understanding the entire business. It depends on a clear understanding of the markets in which it operates, its own strengths and its weaknesses and a clear appreciation of its competitors. The aim is to develop optimum strategies to deliver superior shareholder value. Of necessity, this will bring step changes in competitive performance and ultimately the ability of the business to create sustainable and superior long-term growth in value. Some powerful analytical tools are now available to help quantify different scenarios relatively quickly based on the identification of key value drivers. The clarity of purpose that this can help create for making the right strategic choices can be quite stunning.

Delivering value through integrated performance management

Having decided upon a strategy to create optimum value the next logical management process to focus on is concerned with strategy execution. This is a key element of VBM. History suggests that there is, unfortunately, a very substantial rate of failure in delivering an organisation's vision and strategy.[3]

The key to effective strategy execution is to implement an integrated performance management (IPM) approach. This requires a measurement and information framework that needs to be blended with other vital ingredients of the full value management recipe in order to embed it within the business successfully. A successful IPM approach will help to ensure that strategic intentions are converted into reality in a consistent and comprehensive way. Evidence suggests that a management team's track record and perceived ability in strategy delivery can, in itself, also add significantly to the actual value of the business.[2]

The three perspectives described above are the essential ingredients of a successful VBM approach. Therefore they also provide us with the key themes that underpin the whole of this book.

HOW THIS BOOK IS STRUCTURED

Our central theme is that VBM is all about harnessing, at executive level, these three top management processes in harmony to maximise shareholder value. This is reflected in the structure of this book, as represented in Figure 1.5.

Before looking at the three perspectives in detail, however, we will start by examining the common measurement language used within each of the perspectives. In Chapter 2, we will investigate the metrics that measure the value that companies deliver and investors perceive.

Fig. 1.5 The structure of this book

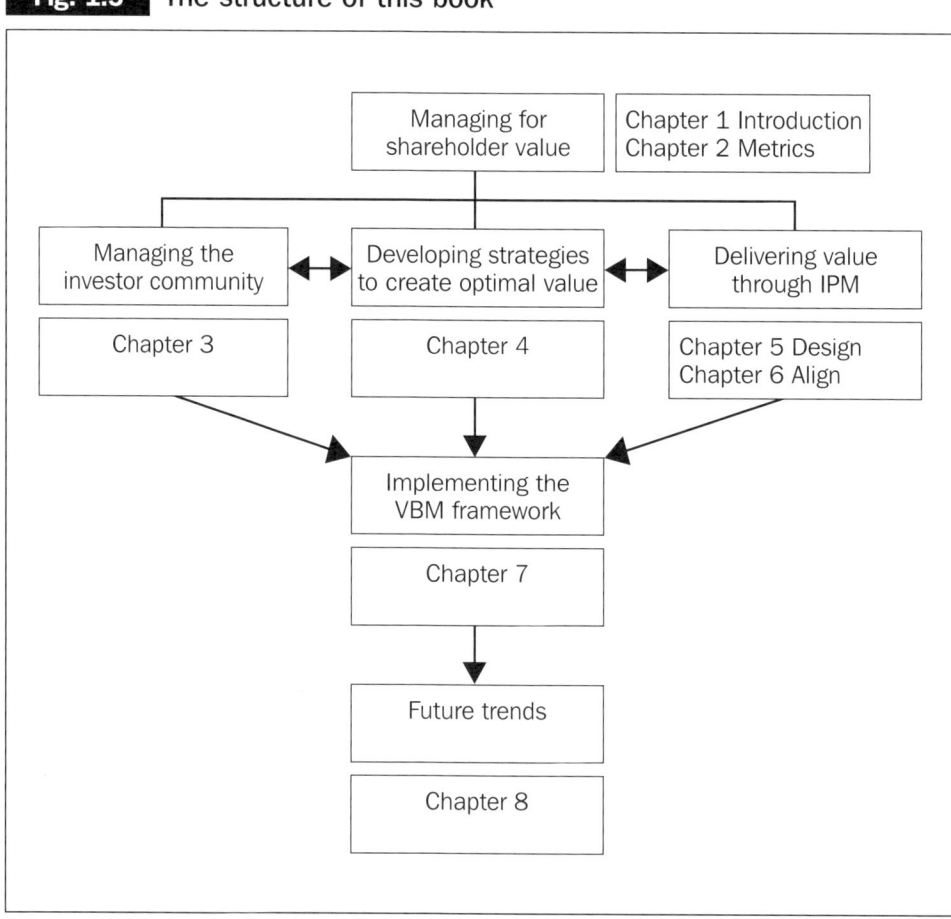

NOTES

1. Martin and Petty (2000) *Value-Based Management, the Corporate Response to the Shareholder Revolution*. Harvard Business School Press.

2. Ernst & Young *Measures that Matter*. Boston: CBI.

3. Thomas A. Stewart (1998) *Fortune*, vol. 138, no. 6, pp. 153–8.

The broad range of shareholder value metrics

CHAPTER SUMMARY

■ Shareholder value metrics are a broad family of techniques and measures. Some are relatively new; others are traditional share valuation techniques.

■ DCF-based valuation measures represent the newer breed of metrics and their use is increasing. They include TSR, TBR, economic profit, CFROI and EV+ among others.

■ TSR is a key metric for investors. It comprises both dividend yield and capital growth, as a percentage of share price.

■ Companies use internal versions of TSR – often called TBR – to cascade TSR into the business and set hurdle rates of return.

■ Economic profit is profit after tax less a financing charge on capital employed. It is a measure of performance which can be extracted from accounting information and which relates closely to shareholder value.

■ With all DCF-based measures, cash is the key to long-term shareholder value. Analysts are increasingly assessing FCF. They project FCF into the future and compare its present value to current share prices. Share prices are a function of expected future performance so this enables the use of more modern DCF valuation techniques such as CFROI and EV+.

■ CFROI is used by some companies and investors as a performance and valuation measure. It overcomes some of the weaknesses of accounting-based measures and seeks to calculate a more valid percentage return on investment in real cash terms.

■ EV+ is an emerging metric that is primarily forward looking. It helps to value shares and sectors by focusing on the analysts' consensus view of the value companies will add in the future. It also identifies possible over- and under-pricing.

■ In addition to these DCF metrics, stock market analysts also use a range of more traditional measures to compare and value businesses. These include P/E ratios, dividend yield, EV/EBITDA and growth measures.

■ It is probably wrong to focus on one school of metrics. Since share prices – and therefore shareholder value – can reflect any or all of these measures, managers should see them all as relevant to varying degrees and should seek to understand the context and meaning of each one.

I wrestle over how to create value, from the time I get up in the morning to the time I go to bed – I even think about it when I'm shaving.

Robert Goizueta, CEO Coca-Cola, 1981–98

METRICS – THE LANGUAGE OF SHAREHOLDER VALUE

In Chapter 1 we introduced the framework of the three perspectives of value-based management: managing the investor community, evaluating strategies to create optimum value and delivering value through IPM. We also referred briefly to TSR and economic profit and we mentioned that these measures provide a common language we can all use when talking about value.

The language of shareholder value consists of a much wider vocabulary than this, however. Before investigating the three perspectives further it is important to understand in more detail the variety of measures used in the language of shareholder value. This chapter will now therefore look further at a broad range of shareholder value metrics that are usually associated with VBM, some of which are relatively new. It will also cover some of the more advanced measures that are currently being employed by the investment community and will review some of the common data items used by analysts to value shares.

Most of the measures covered here are relevant to one or more of the three perspectives. For instance, variants of the economic profit concept are used by the financial community to value shares. They are also used by companies to evaluate alternative strategies and, in addition, are an important part of many company performance measurement systems. This concept is particularly well understood both by companies and by analysts. As a result it provides a common language in communicating shareholder value to the investment community. Companies that show themselves to be using economic profit effectively may, therefore, generate confidence among certain stock market analysts and this in turn can have a positive impact on share price.

Economic profit is a derivation of the principle of 'residual income' that seems to have originated within General Motors and was commonly put forward as an important measurement concept by American business school professors in the 1960s and 1970s. Economic profit became the most commonly used label in the 1980s, but it was primarily when Stern Stewart, the New York based consulting firm, came up with its own trademarked version and gave it the appealing title of EVA (Economic Value Added) that it really began to catch the imagination and the headlines.

However, the introduction of new metrics or new labels does not necessarily justify throwing out all the old measures. Indeed this can be positively dangerous for two reasons. First, it can lead to a rejection by management of the new approach because they feel deprived of their familiar performance measurement systems. Second, it can result in a missed opportunity to build on what is good about the current system. There is a powerful argument, for example, that those companies whose managers are familiar with measures such as profit margin and return on capital employed should leave them in place, at least initially. Through communication, education and training, managers should then be guided towards the new or incremental insights that they bring. The training should emphasise the interaction with other value drivers, such as growth, and show how together they can create increased shareholder value.

It can be argued that analysts today use both DCF-based techniques as well as the more traditional valuation measures when producing price targets for the companies under their scrutiny. As a result, this chapter deals first with the DCF measures and then addresses the more long-established methods.

SHAREHOLDER VALUE AND DCF-BASED METRICS

This section starts with an explanation of some of the main DCF value-based metrics. The measures covered here are:

- *Total shareholder return (TSR)* – which primarily measures historic returns to investors. It is an external, or market-facing, measure.

- *Total business return (TBR)* – which measures the historic returns earned by companies and provides a hurdle rate of return for future performance. It is an internally facing measure.

- *Economic profit* – which measures the surplus profit that companies earn over and above the required hurdle rate of return. This is also called the 'value-added' by companies in a particular period. It is primarily used as a historic, internally facing measure.

- *Free cash flow (FCF)* – this represents the surplus cash a company earns from its operations. This cash is available for future capital investments, for repaying financing or for returning to shareholders in the form of dividends or share buy-backs. It is therefore a key figure in many DCF calculations.

- *Cash flow return on investment (CFROI)* – this shows the percentage rate of return a company is likely to earn in the future. The return calculation is based on the current share price and the cash flows it is likely to generate from the business.

- *Enterprise value added (EV+)* – this is similar to economic profit. It is an absolute measure, not a percentage return. However, it is designed as a forward-looking measure of the value a company is likely to add if it achieves the earnings and cash flows the analysts are forecasting that it will generate.

Total shareholder return

TSR is the normal starting point for most companies that have adopted a value-based approach. It is the starting point because it is the way in which the corporate financial goal is usually expressed.

TSR does not measure return from an internal company perspective but from the point of view of the shareholder – what returns do they receive from holding the company's shares? That is true and real value as seen by the shareholder.

As mentioned in Chapter 1, TSR is the sum of the value received by the shareholder, both from dividend and capital growth, expressed as a percentage of the price of the share at the beginning of the relevant period. Figure 2.1 shows an example TSR calculation for one year.

Fig. 2.1 Example calculation of basic TSR

Share price performance	
■ Opening share price 1 January 2002	100
■ Dividend paid at the end of the year	5
■ Closing share price 31 December 2002	120
TSR calculation	
Dividend yield (dividend paid divided by opening price 5/100)	5%
Percentage price growth (closing price divided by opening price 120/100)	20%
TSR (dividend yield plus percentage price growth 5% + 20%)	25%

For the shareholder this measure represents a means of comparison with other shares and with alternative forms of investment. It is the same approach to evaluation as any one of us might adopt when looking at types of investment. We evaluate the overall return. Thus at one level, TSR is a very simple measure.

There are, however, a number of reasons why it is not as simple as that. For one thing the period of one year is not long enough for a reasonable comparison. Equity investments are by their nature volatile and a share price may be distorted by temporary stock market conditions. The economic environment may have a favourable or adverse impact on all share prices and this may distort the true position of one company in one isolated period.

Therefore TSR tends to be measured in relative terms, compared to other equity investments, and over a period longer than one year. The longer period, though correct and logical, does make the calculation and interpretation of TSR comparisons more difficult. If you look forward, it becomes a very similar calculation to those made by companies when they decide to invest in new projects.

The calculation of TSR over a longer period can be expressed graphically as shown in Figure 2.2. The arrows above the line represent cash or value received by the shareholder at different points in time, or *cash inflows*. The arrow below the line represents the purchase of the share, cash spent by the shareholder in the past, or *cash outflows*.

Fig. 2.2 The TSR model

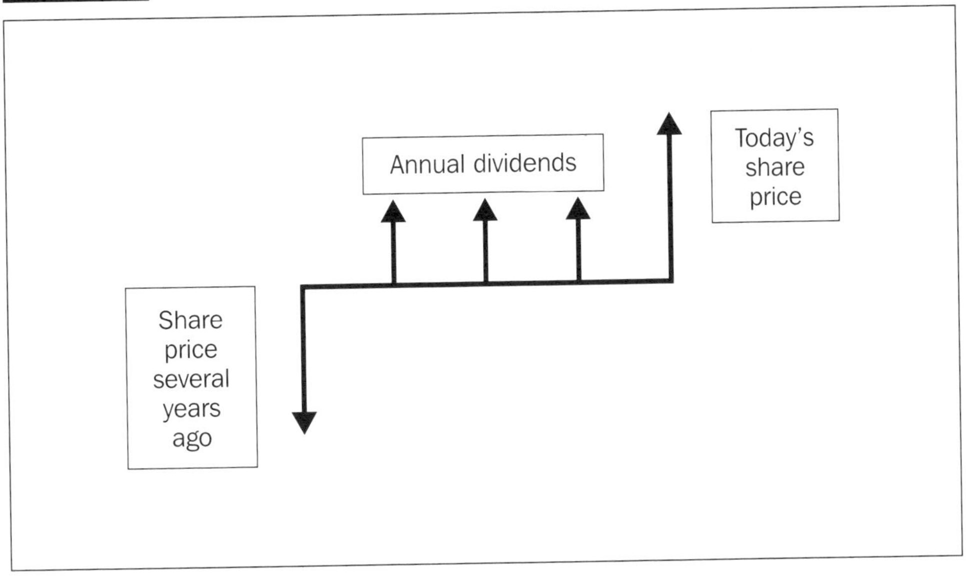

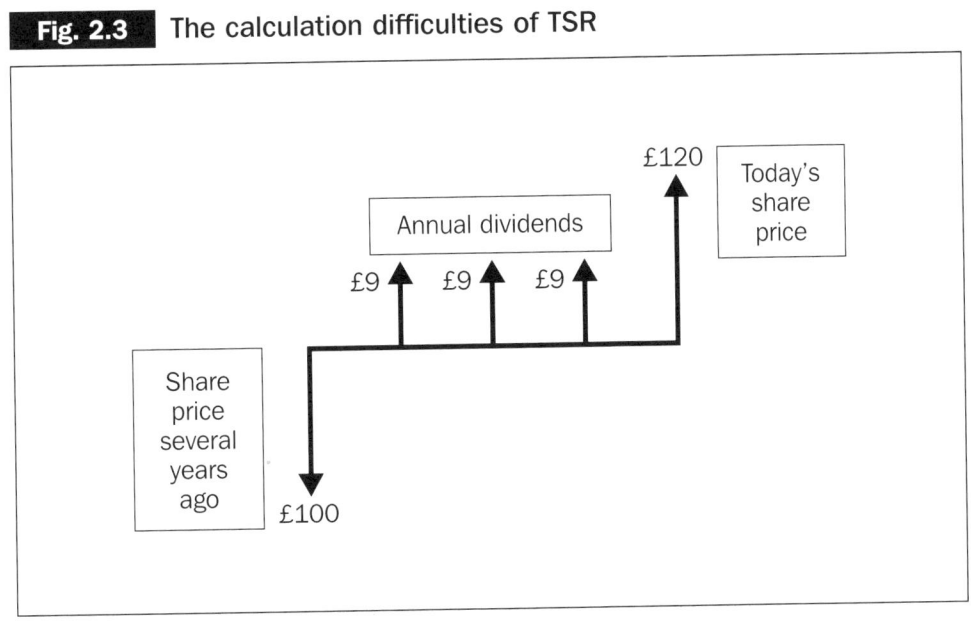

If we add a few simple numbers, we can begin to see the problems of calculating and expressing TSR. Figure 2.3 shows the same model but with some example numbers added.

Fig. 2.3 The calculation difficulties of TSR

The next step is to calculate a TSR percentage over the entire period shown and express it in a way that can be compared with other shares. The shareholder will also want to know whether the return exceeds the cost of the capital required to finance the initial £100 and it is therefore important to have an annual percentage return to compare it with. We will assume that the cost of capital is 10%.

The evaluation of this transaction presents a number of problems. At one level we could say that we have made a capital gain of £20 (£120 sale price less £100 purchase price) and received dividends of a further £27 but that would be a simplistic calculation. To calculate true shareholder value, we need to address some further fundamental questions:

- How do we take account of the time value of money, in particular the fact that the three key elements – purchase price, dividends and sale price – are all in different time periods?

- What assumption do we make about reinvestment of the dividends received during the period?

- If we make some kind of percentage return or present value calculation, how do we express it, what do we relate it to?

Today, software packages make it simple to calculate a TSR from a series of cash flows. For example, any standard spreadsheet will quickly calculate that the TSR for the above transaction is a little over 11%. The method of arriving at this number will be familiar. It is the same DCF methodology that arrives at an internal rate of return for the future cash flows of investment projects, as explained in Appendix 1.

Shareholders can use historic TSR to help answer questions such as:

- What level of TSR has this company achieved in the past few years?

- How do these TSRs compare with those of other companies?

- Does the excess over the cost of capital (in this case 10%) compensate us for the particular risk we are taking by buying this share compared to other less risky (or risk-free) alternatives?

TSR does not need to be restricted solely to the measurement of historic performance. If used to aid investment decisions the model needs to be adapted to incorporate the future performance of the share. While this is easy to describe, it is not easy to do since it involves predicting the growth in share price in the future. Shareholders can then use TSR to help with investment decisions by answering questions such as: 'What TSR will this particular share achieve if we buy at today's price and the forecasts we are making about dividend and capital growth are achieved?'

Above all, whether focused on the past or the future, TSR provides a practical means of comparing the overall performance of all the shares in the market over a given period. The results are highly dependent on the point of time at which the calculation is made (though this applies to any measure derived from volatile market values) so care should be taken to choose a time that is not too distorted by short-term market fluctuations. It is also important that the results are seen in comparative terms over a long period. Three to five years would be a typical timescale.

This is precisely how some professionals use TSR. They produce lists of comparative TSR performance based on past dividends and share prices. Figure 2.4 shows a typical comparison made by the Boston Consulting Group of top shares in the consumer goods sector over a ten-year period.

Fig. 2.4 Comparative TSR performance

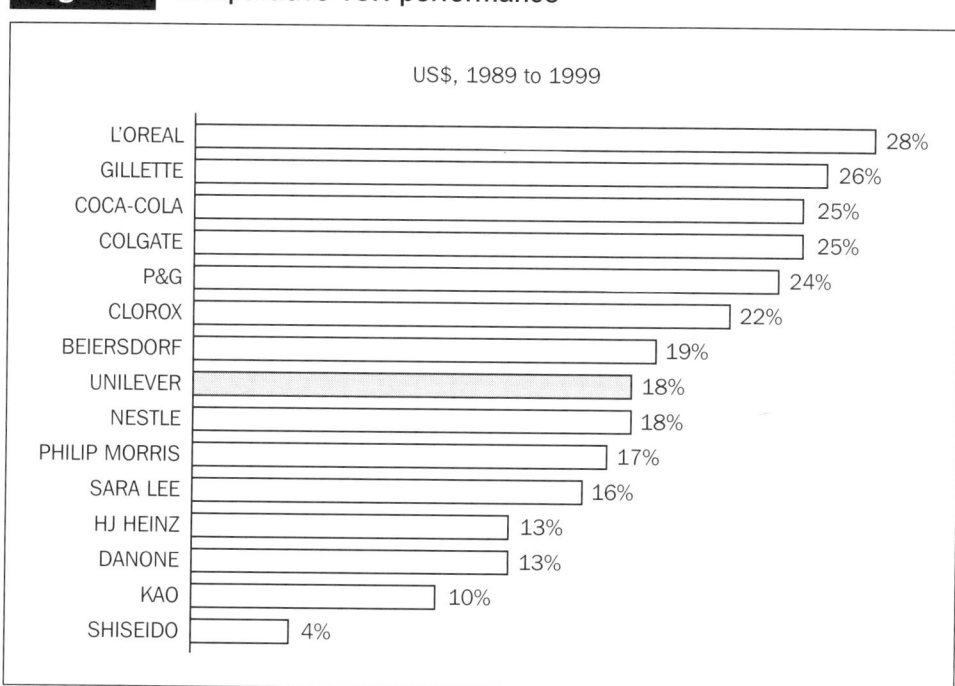

Source: Thomson Financial Datastream

The TSR measure may seem in many ways to be an obvious choice for shareholders and therefore for the top management who are there to serve them. Yet only relatively recently, and even in some of the world's most well regarded companies, the move to TSR as the primary measure of performance at corporate level was regarded as something of an event and a breakthrough.

The following section examines one of these companies, Unilever, and describes the path they have taken towards TSR.

CASE STUDY – UNILEVER'S ADOPTION OF TSR

In October 1997, the Joint Chairmen of Unilever formally announced to the financial community that the company was no longer taking earnings per share and other accounting based measures as main indicators of its total company performance. They were to be replaced by TSR, which became the single most important measure for the main board. The statement to analysts said:

> *TSR is the best single measure of the inherent value creation of a business. It needs to be measured over a sufficiently long time; we have chosen a three-year rolling period. As TSR performance can only be judged relative to other companies, we have selected a peer group of 20 FMCG [fast moving consumer goods] companies, each international, each with sales of over $1bn.*

Unilever is not the first company to do this and will certainly not be the last. What was particularly interesting and unusual was the public and positive way in which this change was announced. This reflected the increasing importance that top management and the financial community attached to having the right shareholder value measures in place.

However, once a company has adopted TSR as a corporate goal, the next step is to adopt an appropriate set of internal performance measures that encourage managers to achieve the new targets.

Despite all its advantages, and despite its widespread adoption as a key value-based measure, TSR is not easy to cascade down into the business. Divisional managers and their staff are likely to find it difficult to be motivated by such a measure and a key goal for major companies has been to find a way of capturing TSR's key elements in other measures that can be used flexibly within the business. This simple goal – finding the best link between TSR and internal metrics – has given rise to a number of other value-based measures that vary in their complexity and validity.

The first to be discussed is TBR and then the link to economic profit will be explored.

TBR – an internal hurdle rate of return

A further statement was made by Unilever to stock market analysts on the announcement of TSR as its main corporate financial objective. This extract concerns TBR:

> *We have developed an internal measure which mirrors TSR, but which can be used within Unilever. We call this Total Business Return or TBR. It will align the targets for management throughout Unilever with the corporate target and change the behaviour of our managers.*

TBR has its variants. There are similar models with different names but the principles remain the same. Major companies using this approach ask their business units to convert their long-term plans into cash flows, in the same way as they might prepare a forecast cash flow for a major investment project. The cash flows are then converted to present values and a total business return (i.e. internal rate of return) is calculated in exactly the same way as with the TSR example we showed earlier. The time period used for TBR is a question of judgement, based on the company's business cycle and the features of the business sector.

This sounds easy in principle but in practice it is not easy at all. Yet it is an important process if TSR is to be cascaded down effectively into business units. The idea is that the plans of each business unit are evaluated as a TBR percentage and that collectively the agreed targets match the company TSR. The TBR model is shown in Figure 2.5.

Fig. 2.5 The TBR model

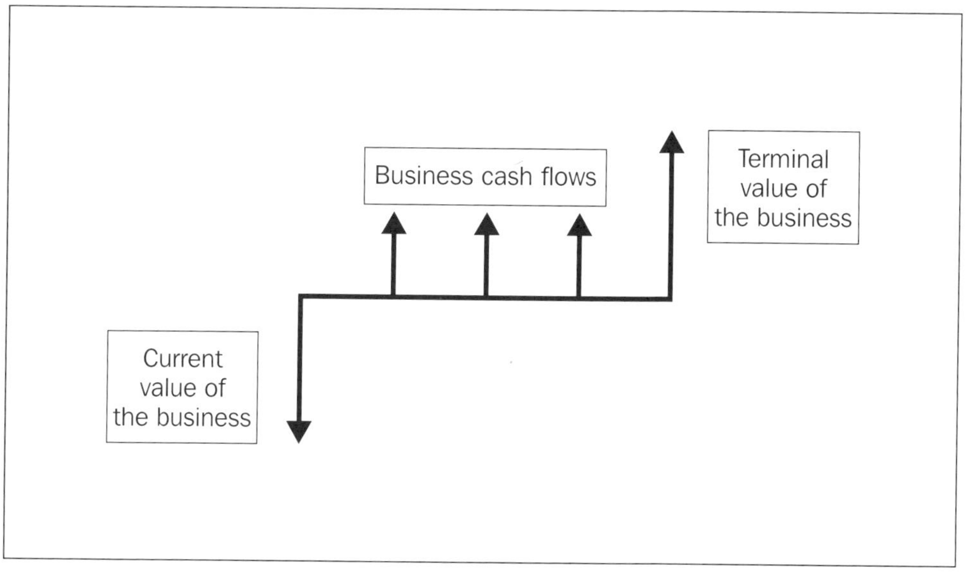

Note the similarity of TBR to the TSR model. The cash flow generated from a business unit during the relevant time period is the equivalent of TSR's annual cash dividends. It is the net amount of cash delivered to the centre by that business unit during the period.

The calculation of current and terminal values for each business unit is, however, problematic. These are the TBR equivalents of TSR's opening and closing share prices. This calculation is not easy and can get very complicated. There are a number of possible approaches. None is perfect and all are likely to be challenged. At the simplest level a business unit may be valued by using an estimated earnings multiple. At the other extreme there are very complex approaches that involve the use of mathematical formulae. The final choice is a matter of judgement but the objective is the same – to arrive at a meaningful and realistic value of the business unit, *as it would appear if quoted in the stock market.*

There are reasons why technical perfection in this area is less important than is often thought. There is never a single answer to the valuation of a business, any more than there is a single answer to the valuation of a car or a house. Provided the same method is applied to both the current and terminal values, it is not always necessary to become too concerned about perfection. The important things are the *changes* to the factors that drive value, not the absolute numbers themselves.

The TBR framework is most valid for long-term planning and analysis purposes. Like all cash flow measures and models, it does not provide an easy way of measuring performance in one particular year. What it does do very powerfully is to provide the framework for setting operational performance targets arising from the long-term plan.

From a strategic plan based on TBR, targets should flow that relate to the main drivers of value creation. We will cover this in more detail in Chapter 4. The problem then is how to measure the effectiveness of each target using one simple metric that reflects them all in any given accounting period. There can of course be separate metrics for each but it is very helpful to employ one measure through which senior managers can monitor their performance against plan. This is where economic profit fits in.

Economic profit

Economic profit is a technique for measuring the value created in a single period. It fits neatly within the frameworks for TSR and TBR and uses some of the same principles they employ.

Economic Profit represents the difference between the *sales* made in a period and the *real, up-to-date, cost of all resources* consumed in that same period.

Some economic profit approaches seek to overcome the weaknesses of conventional accounting by valuing resources consumed at current, rather than historic, cost.

It is, however, in its treatment of the economic cost of using assets within the business that economic profit is significantly different from conventional accounting measures of profitability. The key feature of economic profit is that it brings balance sheet and cash flow variables into the profit and loss account. It does this by charging, as an additional expense against profit, the company's cost of capital as a percentage of assets employed in the business. So:

Economic profit = Post-tax profit less a charge on capital employed

Post-tax profit is not as easy to extract as it may seem and this is one of the practical complexities of arriving at a true economic profit number. The normal accounting definition of post-tax profit, the earnings number in the profit and loss account, is after the deduction of interest costs. However, the charge on capital employed also includes the cost of interest so an adjustment has to be made to avoid double counting. The actual interest cost is therefore added back and the tax charge has to be recalculated based on the revised profit figure. The resulting number is often called NOPAT – net operating profit after tax.

The charge on capital employed is calculated as follows:

$$\text{Capital employed} \times \text{Weighted average cost of capital}$$

The weighted average cost of capital (WACC) is the average cost of debt and equity capital. The calculation is weighted according to the proportions of the two types of finance and is normally calculated in the same way as the cost of capital for investment projects. The precise calculation of WACC is shown in Appendix 2 of this book.

Figure 2.6 shows a simple example of an economic profit calculation, based upon the above concepts.

Fig. 2.6 Example of economic profit calculation

Financial statistics	
	£
Capital employed	100
Post-tax profit (NOPAT)	30
Weighted average cost of capital	10%
Economic profit calculation	
	£
Post-tax profit (NOPAT)	30
Less charge for capital employed (£100 × 10%)	(10)
Economic profit	£20

The simplest way to understand the meaning of this economic profit number is that it represents what is left for shareholders after all costs have been covered, including the cost of borrowing and their own required return. A positive number means that value has been created; a negative number means that value has been destroyed.

An alternative way of calculating economic profit is to use what is known as the 'spread' method. This involves taking the excess of the return on capital employed (ROCE) over the cost of capital and applying it to the capital employed number. It should produce a result that can be reconciled to the approach above. It has the advantage of relating economic profit more closely to ROCE and also makes it clear that there is no real return until the cost of capital has been covered.

Using this method the calculation of the same answer would be derived from the formula:

$$\text{Economic profit} = (\text{ROCE} - \text{Weighted average cost of capital}) \times \text{Capital employed}$$

Using the figures above the ROCE is 30% (£30 post-tax profit as a percentage of £100 capital employed) less 10% WACC, leaving a 20% spread. When applied to the capital employed of £100, this gives an economic profit of £20.

Another way of thinking about this 20% spread is that it is the 'economic ROCE' or 'economic profit spread', the return being made in excess of the cost of capital. Some companies use it in this way and it can be a helpful indicator, though it still suffers from the same limitation as any percentage measure. It does not encourage maximisation of value in absolute money terms.

Stern Stewart has identified over 150 possible adjustments which can be applied to the profit or capital employed figures before computing their version of economic profit, EVA. Some of these are described as behavioural, reflecting a desire to influence management actions in certain ways, depending on strategic priorities. This confirms that the calculation of economic profit is, to say the least, an inexact science. Appendix 3 contains more information about some of the most common adjustments companies can and do make in calculating economic profit.

Therefore, for companies to use economic profit as a true measure of shareholder value, they must do more than merely deduct in their profit and loss accounts an additional charge for cost of capital. Many companies use so-called economic profit metrics that do no more than make this one adjustment. For those companies a fundamental problem still remains because all the numbers above that line are still based on accounting concepts. Many of these concepts are contrary to our earlier definition of economic profit and do not reflect the type of adjustments highlighted in Appendix 3.

Often the main focus of these adjustments is to change the timing of expenses. So items such as research and development, which might normally be recorded as expenses in the period in which they are incurred, can be capitalised as 'intangible' assets that will then be depreciated over a number of periods, their cost being recorded against the revenues they will subsequently generate.

By shifting expenses in this way, and by allowing some of the conventional accounting adjustments, economic profit methods can be said to depart from straightforward cash flows. As such they are at odds with the more standard DCF principles embodied within TSR and TBR.

This raises an interesting issue. While economic profit can provide a measure of the value-added by a company in a single period, it does not always easily reconcile with some of the cash-based share valuation techniques being employed by analysts today. This is because accounting profits, and therefore economic profit, can be manipulated. Many analysts are, therefore, now adopting valuation techniques that use a company's cash flows, rather than its profits.

Free cash flow

One of the main measures of cash flow is known as 'free cash flow'. This, too, must be viewed as an important shareholder measure since it often forms an important part of analysts' share valuations.

The key objective of free cash flow assessment by analysts is to project forward into the future as far as it can be forecast. This is because, for a share price to be sustainable, those buying the shares must believe that the company will be able to generate enough future cash flows to justify that price. If analysts find that the share price is worth less than the present value of future cash flows, they may well recommend investors to sell the share and the price will fall.

The DCF model, as discussed in Chapter 1, applies well to help value shares. It contends that the 'value' of a share is equal to the PV of the future dividend stream to the investor. Since some companies do not pay dividends, and tax can affect dividend distributions, academics argue that cash flows or earnings can be substituted for the dividend stream and the result should be the same. So the PV of the anticipated dividend (or cash or earnings) stream, less the cost of the share today, represents a view on the over/underpricing of the share.

Today, partly because of the theoretically sound basis of the DCF model, there is an increasing refocus on cash flow as a key driver of share price. Earnings-based valuations can be distorted by numerous accounting adjustments and are also subject to varying accounting standards in different countries. This makes it possible for companies to manipulate their reported results in order to influence their market valuation.

The link between free cash flow and TSR is important. The analysts who use TSR, and the companies who have made it their main corporate goal, have realised that there must be a relationship between the ability to deliver TSR and the free cash flow being generated by the company. They are both cash flow measures. TSR is the return to shareholders; free cash flow is the cash surplus generated internally by the company's operations, before allowing for spend on reinvestment or financing. Thus analysts who want to find companies who will deliver good TSR in the long term will look for companies who will generate high levels of cash, a further argument for the focus on free cash flow over time.

However, free cash flow is not as simple as it might first appear. This is because the definitions of free cash flow vary, and the detailed methods of arriving at this number will differ. However, the objective is usually the same: to find out how much cash the company is generating that is potentially available for shareholders. Not all companies use their surplus cash to pay dividends, so by valuing the surplus cash, rather than the dividend stream itself, analysts theoretically produce a purer valuation. Free cash flow information is usually available from published accounts though not all companies make it easy for the relevant numbers to be extracted.

A typical route to free cash flow is through the 'indirect' method, starting with profit and working back to cash. This method, widely adopted by US analysts, is explained in Figure 2.7, with some simple numbers to show the relationships.

Fig. 2.7 Example of free cash flow

Earnings (after interest and tax)	100
Add back depreciation and other non-cash items	25
	125
Increase in working capital (stock, debtors, creditors)	(50)
	75
Capital expenditure	(60)
Free cash flow	15

Some of the definitional problems of free cash flow relate to the capital expenditure element. As mentioned above, it is not possible to draw too many conclusions from one year because capital expenditure needs are likely to be volatile from year to year in many cases. As a result some analysts take a 'normalised' figure for capital expenditure, trying to arrive at a typical annual spend over the long term. Others include only that part of capital expenditure that is required to maintain normal operations, excluding that portion which effectively provides for future expansion.

Another grey area is the question of whether acquisitions should be deducted, as their impact is effectively the same as that of capital expenditure. The answer will depend on whether acquisitions are likely to be a regular part of the long-term strategy.

There are more advanced approaches to cash flow analysis. Some analysts and investors attempt to value the 'real options' embedded within a business. They use complex techniques that value the cash flows that may arise if companies take certain investment decisions in the future. These techniques recognise that companies face *different business choices* at *different points in time* and that they can choose to invest in all or part of any particular project when they know enough about it. They do not always have to make 'all or nothing' decisions. For example, oil companies will delay their decisions to develop new wells in a particular site until after they have invested in initial explorations of the area.

The management of a quoted company should constantly be making its own cash projections and trying to simulate those of the analysts, so that comparisons can be made to actual stock market value. Having identified any gaps between future cash projections and stock market value the challenge is to fill them or explain them away. Either way, it is vital that the company is aware of these gaps. The options for action will be explored in Chapters 3 and 4.

We will now investigate two of the more modern metrics that employ the DCF and value-added techniques described earlier in this chapter. The first of these is CFROI.

Cash Flow Return on Investment

CFROI™ is a trademark of HOLT Value Associates, LP and is, in itself, the subject of a 356 page book (Madden, 1999). This demonstrates some of the comprehensive nature of the metric involved. It is a performance metric that is used by a number of investment institutions to help value companies. It is a DCF measure which is designed to aid the understanding of share prices on a global basis.

CFROI is a period performance measure, based on future long-term cash flows, which includes all the accounting, inflationary and economic adjustments necessary to allow direct comparisons over time, across industry sectors and international borders. The calculation is again based on the internal rate of return approach described earlier.

In this valuation framework, the business is viewed as a portfolio of projects. The profit and loss account in any one year represents the sum of the funds generated by each project in the business. Some of these projects may be just beginning; others may be mature or closing. So, in any particular accounting period, for one project it might be the first year of cash generation, for another it might be the last. The model also assumes that the balance sheet provides information about these investments and that funds will be released from the investment assets at the end of each project.

Clearly, the full CFROI calculation is too complex to cover fully in this book but in essence it has the following four steps:

- Adjust published cash flow and investment figures to give an economic perspective which excludes inflation and other distortions.

- Estimate the lives of the different classes of assets and use this to project the cash flows they generate.

- Estimate the likely value which will be released when these depreciating assets reach the end of their useful lives.

- Calculate an internal rate of return for the business. This is the CFROI.

Having calculated CFROI the next step is to project the future cash flows of the business, allowing for growth estimates and the fade of any competitive advantage. These are then discounted at an assumed cost of capital to provide an estimate of enterprise value.

From the above description it can be seen that a large part of the CFROI valuation framework is devoted to the creation of the forecast performance that is then used for the internal rate of return (CFROI) calculation.

The next section covers an emerging value-based measure that uses the forecasts of analysts as input to the calculation, instead of producing its own performance estimates. This measure is known as Enterprise Value Added.

Enterprise Value Added

Enterprise Value Added (EV+) is essentially a DCF-based residual income technique, produced and patented by Sharevaluer. It incorporates many of the newer cash-based approaches found in modern analyst reports. In some respects it is calculated in a similar way to economic profit. However, EV+ is different in two key ways:

- First, EV+ focuses solely on the future, not the past. EV+ is based upon the forecast profits and cash flows produced by analysts, which are the key determinants of share price and the most important variables in a DCF share valuation.

- Second, EV+ uses the increasingly popular enterprise value (EV) as the basis for the capital charge. EV is the *market* value of both debt and equity capital so it represents the value of the company as perceived by investors. The use of EV in the calculation makes EV+ a technique which measures returns on 'true' business value. Other economic profit metrics base their cost of capital charge on historic balance sheet values, which are not always relevant to investors or reflective of the true value of the business. This makes them essentially internally facing.

Being a residual income technique, EV+ measures the 'value added' to shareholders in each of the years within the forecast horizon. Over the long term, if these values are negative, the share is technically overpriced. If positive, the share is underpriced. Because EV+ identifies the components and timing of share mispricings it is of direct relevance to investors and analysts.

EV+ effectively produces a DCF-based target value for the company and aligns with forecasts of TSR, which comprises target value plus forecast dividend yield. This enables investors to choose between alternative investments based upon the TSR they are *likely* to deliver, as opposed to the economic profit they have delivered in the past but may fail to reproduce in the future.

As a result it can be seen that EV+ is not one single value-based measure. It is actually a model that produces a whole family of different metrics that support modern valuation principles, measuring returns against the true 'enterprise value' of the company. An example of EV+ and its related metrics is included in Appendix 4.

Because DCF-based measures, such as CFROI and EV+, are becoming more widely used by investors and analysts, they are also likely to become of importance to corporates who seek to demonstrate the shareholder value they plan to add over the foreseeable future.

However, it would be totally wrong to imply that value can *only* be defined in DCF terms. Share values have been determined for many years using more traditional approaches. These methods are still in widespread use today. As a result they are equally important for those companies who wish to understand more fully how their shares have been valued. We will now therefore examine the more common ratios that have been, and are still, used to value shares in the stock markets.

SHAREHOLDER VALUE AND NON-DCF METRICS

In practice, investors recognise that it is not correct to value all companies in the same way. Property company values, for example, are largely a function of the value of their underlying assets – the land and buildings they mean to develop and sell. So shareholder value must be measured by reference to the market values of these assets.

There are also many other valuation and measurement techniques that are not strictly based on the DCF methods we have described above. However, they are used so commonly by analysts that no company can afford to ignore them. As a result companies should adopt a broad view of 'value' and seek to understand a whole range of measures that are used within the investment community.

Whereas most DCF measures and valuations are 'absolute', in the sense that their value can be calculated without reference to the valuations of similar companies, most non-DCF measures are not. They are *relative* to the sector in which the company does business, to its competitors or its peer group. As a result a key focus behind the development of newer metrics is to aid comparability between different companies. The main measures used in analysts' reports for comparison and relative valuations are discussed in the sections which follow. They are:

■ *Return on capital employed (ROCE)* – which measures profitability in relation to the amount of capital within the business.

■ *Earnings per share, price earnings and forward P/E* – which measure profitability per share in issue and also in relation to the price of the share.

■ *Dividends per share and dividend yield* – which measure the dividends per share in issue and also in relation to the share price.

- *EV and EBITDA* – which represent the market value of the entire business and a profit metric that enables cross-border and cross-business comparison.

- *Growth and PEG* – which measure the rate at which the profitability of a business is likely to increase.

Return on capital employed

Return on capital employed (ROCE) is a well-known measure of profitability. Because it is expressed as a percentage it theoretically enables the comparison of companies of a different size. The basic formula for ROCE is a follows:

$$\frac{\text{Profit before interest}}{\text{Capital employed (total assets – current liabilities)}} \times 100$$

Because the figure for capital employed is derived from historic balance sheet values it can be distorted. One company may not adopt the same valuation principles as another. Fixed assets are probably not included at their true value and intangible items, such as brands or goodwill, may not be valued at all. As a result, the usefulness of ROCE for comparability purposes can be diminished.

Earnings per share, price earnings and forward P/E

Earnings per share, or EPS, is still the most quoted and widely used shareholder measure. Its calculation is simple. The formula is:

$$\frac{\text{Earnings}}{\text{Average shares in issue during the year}}$$

This ratio shows the amount of profit delivered to a shareholder with one share, either in dividend or as retained profits. A trend can be extrapolated to highlight whether EPS is increasing or decreasing. However, it means nothing in comparison to other companies because the number of shares is a unique characteristic of each company, based on its history and its capital structure.

The pressure from the stock market is to grow the EPS number year on year, in relation to either growth achieved in the past or to that delivered by other companies. There is no doubt that EPS has been a key factor in driving share prices in the past and was traditionally seen as the measure which stock exchange analysts looked at before all others. It is only in recent times that research has questioned the correlation to long-term shareholder value.

To overcome the lack of comparability of EPS figures, the price earnings ratio (or P/E multiple) is perhaps the most widely quoted *comparative* measure on the Stock Exchange. It is calculated like this:

$$\frac{\text{Current share price}}{\text{Earnings per share}}$$

P/E measures the extent to which today's share price, which reflects all available information and future expectations, is a multiple of the most recent year's EPS. The higher the P/E, the stronger are the future expectations for the company. This means that the market expects the company to get better in relation to its current EPS performance. This may show confidence in the quality of management; on the other hand it may merely reflect the poor quality of present performance.

Like P/E, forward P/E is calculated by reference to the current share price but it uses the *forecast* EPS instead of the most recent declared result. It can therefore be calculated for any year in which forecasts are available. This enables comparisons with similar companies, not just for one particular year but for several. For companies that are forecast to grow fast the forward P/E will drop faster than for other companies. So the relative trend in forward P/E, from one company to another, can also be a useful indicator of investor confidence. An example of forward P/E is shown in Figure 2.8.

Fig. 2.8   **Example of forward P/E**

Companies A and B both have the same share price, £2, but different P/E rations. The last EPS of company A was 10p, so it has a P/E of 20. Company B's EPS was only 5p so it has a P/E of 40. Yet company B may not be overvalued in relation to A. Why? The answer lies in the forecast EPS growth pattern of both companies.

	Last Year	Year 1	Year 2	Year 3	Year 4
Company A					
■ EPS last year and 4 forecasts (p)	10	10.5	11.2	11.6	12.2
■ Forward P/E (£2/EPS)	20	19.0	17.9	17.2	16.4
Company B					
■ EPS last year and 4 forecasts (p)	5	6.2	7.8	9.8	12.2
■ Forward P/E (£2/EPS)	40	32.2	25.6	20.4	16.4

B is not overvalued because the market expects it to grow faster than A and this is reflected in the way that B's forward P/E drops faster than A's.

Dividends per share and dividend yield

Just as with earnings one can calculate an EPS, it is also possible to calculate the ratio of dividends per share (DPS). Like EPS, DPS can provide some useful trend information but it is not good for comparing one company with another.

Once again, by recalculating the ratio as a percentage of share price, it is possible to produce a measure that facilitates comparability. Dividend yield is the comparison of DPS to share price, showing the percentage return shareholders receive by way of dividend in relation to current market value.

Again, like EPS and P/E, these ratios can be calculated using historic and/or forecast data.

EV and EBITDA

EBITDA stands for 'earnings before interest, tax, depreciation and amortisation'. It is a useful measure of a company's performance because it enables improved comparability between different businesses and has a closer relationship to cash flow than other accounting-based measures. It adds back to profit a number of items that would otherwise reduce the comparability between different companies. These are:

- *interest* – which varies according to a company's capital structure;

- *tax* – which varies according to a company's home tax regime;

- *depreciation and amortisation* – both of which are subjective non-cash adjustments to profit.

EBITDA can also be expressed as a percentage of sales, normally called the EBITDA margin. Again this measure enables better comparisons to be made between companies.

As mentioned earlier in this chapter, enterprise value, or 'EV', is a calculation that arrives at the *full market value of the business, irrespective of its capital structure*. It comprises the current market capitalisation of shares plus the market value of the company's debt. Some advocates of EV also make various adjustments for non-operating assets in order to arrive at a market value of the operating business.

Because EV is a measure based on market values it is directly comparable across companies. So a company with an EV that is greater than another can truly said to be worth more, in exactly the same way as share prices can be compared between companies. However, EV on its own, does not attempt to measure performance. For that analysts use EV/EBITDA.

Both EV and EBITDA take on more meaning when they are combined together as the ratio EV/EBITDA. This measures the relationship between the value of the company and its earnings, as expressed by EBITDA.

EV valuations are often expressed as multiples of EBITDA to enable comparisons of valuations of businesses in the *same sector*. Other EV-based ratios, for example EV/Sales, are also frequently found in stock market analysts' reports to enable comparison with other businesses of a different size.

Growth and PEG

Growth has recently become an even more important factor in measurement and valuation. Since accounting-based ratios, such as EPS and P/E, tend to measure results in a single period, they frequently ignore the likely future growth prospects of the company. Therefore analysts attempt to measure the extent to which some companies grow faster than others. Clearly, although two companies may have the same EPS today, if one is expected to grow its EPS faster than the other then it will be worth more.

One of the main growth measures is EPS growth. This can be measured over a single year or averaged over several years. In addition analysts sometimes produce estimates of longer-term growth rates, to represent the anticipated growth in earnings over the next business cycle of the company.

The price earnings growth ratio (PEG) uses growth by incorporating it into the standard P/E ratio. PEG is a ratio that enables comparison of the rating of different companies in relation to forecast growth. It is obtained by dividing the P/E ratio by the forecast percentage growth in EPS. If two companies have the same P/E, the one with the higher growth rate will have the lower PEG. Therefore investors who seek growth will invest in companies with relatively low PEG ratios. An example of PEG is shown in Figure 2.9.

Fig. 2.9 **Example of PEG**

Consider again companies A and B from Figure 2.8. They have the same share prices and EPS forecasts as before.

	Last Year	Year 1	Year 2	Year 3	Year 4
Company A					
■ EPS last year and 4 forecasts (p)	10	10.5	11.2	11.6	12.2
■ Forward P/E (£2/EPS)	20	19.0	17.9	17.2	16.4
■ Forecast EPS growth (e.g. Yr 2/Yr 1)	N/A	5.0%	6.7%	3.4%	5.2%
■ PEG (Forward P/E divided by growth)	N/A	3.8	2.7	5.1	3.2
Company B					
■ EPS last year and 4 forecasts (p)	5	6.2	7.8	9.8	12.2
■ Forward P/E (£2/EPS)	40	32.3	25.6	20.4	16.4
■ Forecast EPS growth (e.g. Yr 2/Yr 1)	N/A	24.0%	25.8%	25.6%	24.5%
■ PEG (Forward P/E divided by growth)	N/A	1.3	1.0	0.8	0.7

B has significantly lower PEG figures than A because it shows greater growth potential. It therefore justifies its currently high P/E.

THE MEANS OF COMPARING COMPANIES

Sector aggregates are used in almost all analysts' research. These are the average results for the entire group of companies within a given sector. Sector aggregates are normally computed for all the main corporate measures we have mentioned above, enabling a comparison of the rating of one company with the mean of its entire sector. This will enable analysts to determine whether shares are currently trading at a premium or at a discount in relation to the rest of the industry.

Peer group and competitor comparisons are an advance on simple sector aggregates by the calculation of averages for selected groups of similar companies. These improve on sector comparisons because they can be restricted to specific companies, for example those of a similar market capitalisation within a particular industry. They can also be used to identify the mean for a company's known competitor peer group. Examples of the use of these peer groups by Barclays and BP, were given in Chapter 1.

WHAT DO ALL THESE MEASURES TELL US?

In relation to the DCF measures we covered in the first part of this chapter it could be argued that the real message of value-based management is that, in the long term, 'cash is king'.

The trend towards VBM and shareholder value metrics could be seen as a move away from the accounting-based measures we covered in the second half of this chapter. The move could be argued as being towards the realities of cash generation. If it really were that simple, of course, there would be no need for all these new metrics but the underlying principle of 'cash is king' should not be forgotten.

The real driving force for change has been the fact that certain stock market analysts began to have reservations about the usefulness of conventional accounting measures. During the late 1980s and early 1990s these opinion formers became increasingly concerned that the management of many companies did not have the same view. Despite general statements of the need to generate cash and create shareholder value, earnings per share and return on capital employed continued to be regarded as the key measures by many management teams. Cash, and the return these companies were providing to shareholders, still seemed to take second place.

Against this had to be set the fundamental problem that cash is *not* a good measure of shareholder value or management performance in the *short term*. It has to be looked at over years rather than months because businesses invest cash for future benefit. Managers and analysts also need to measure success in the short term and therefore look for measures that enable this. This is one of the strongest arguments in favour of economic profit.

However, companies and analysts also need to carry out longer-term analysis of cash flows, particularly into the future because that is the key to value. The reason for this is simple and goes right back to an important premise: *at any one time a business is worth the present value of its future cash flows*. Such analysis is increasingly being carried out by stock market analysts and, to do so, they need to arrive at the cash that is likely to be generated for shareholders.

This chapter has focused on the new measures that are associated with VBM and on other ratios and measures used by analysts to value and compare companies. These measures should be viewed as a portfolio, a set of tools to be brought out at appropriate times by different players for particular purposes.

Since investment analysts use *all* of these tools to value shares it is important that companies who seek to deliver superior shareholder value appreciate the potential impact that they may have upon share price.

Having examined the metrics that support shareholder value we have already begun to explore some of the key elements of the first of our three perspectives. This covers the need to convey financial information and business performance to the investment community. The following chapter examines some of the other key attributes of that community.

REFERENCE

Madden, B. J. (1999) *CFROI Valuation*. Butterworth-Heinemann.

3

How to convey value
to investors

CHAPTER SUMMARY

- Delivering shareholder value is fundamentally about maintaining or increasing a company's total shareholder return, of which the main component is share price performance.

- The *external* investment community sets share prices.

- It is therefore vital to understand the key players in that community, their perspectives on shareholder value and the information that flows between them.

- A good appreciation of the complexities of the investment community underpins successful investor communications and can help boost share price.

- The key decision-makers within the investment community are the fund managers.

- The main information used by fund managers is contributed by data providers, journalists, analysts and the companies themselves.

- The major data providers offer complex sets of news and data items that are critical to the calculation of shareholder value.

- Analyst reports employ differing share valuation techniques that make use of a wide variety of data items.

- Successful investor relations experts recognise the complexities within the investor community. They actively monitor and manage the perceptions about their companies.

If you are working hard on shareholder value then I should see it reflected in your share price relative to your peer group.

Michael Bamforth, Thomson Financial/Carson

UNDERSTANDING SHAREHOLDER VALUE – THE INVESTOR'S PERSPECTIVE

We have already examined many of the current methods used to measure and drive shareholder value within companies. We concluded that shareholder value measures should form the basis of a common language between a company and its investors. However, the language is somewhat complex.

The previous chapter showed us some of the complexities of share valuation and highlighted some of the differing approaches used by professionals within the stock markets, even when they are basing their valuations on DCF techniques. We included these descriptions to illustrate the need for care when communicating with the investment community. Companies should always attempt to understand the range of specific techniques that analysts are using to value their shares and those of their competitors. This will leave them better placed to identify the buttons they need to press in order to manage the expectations of the investment community as a whole and demonstrate the shareholder value they are planning to deliver.

For companies that want to deliver superior shareholder value, it is vital that they first attempt to understand what the term shareholder value means to them and, most importantly, clarify the major information drivers that affect their share price. We have already discussed the concepts of shareholder value earlier in this book. However, to *really* appreciate shareholder value, companies need to put themselves in the shoes of their investors.

Once companies appreciate the information available to the investment community, and the uses that can be made of it in order to value their shares, then they are much better placed to establish the likely impact of their own results on the value their investors perceive and attribute to their business. This can help them in two key ways:

- First, they can develop strategic targets for their performance which will link more clearly to the shareholder value they will deliver to investors. This will be covered in Chapter 4, when the use of value-based metrics in strategy evaluation are discussed.

- Second, they can improve their communications with investors. The messages they give about the company have a direct impact on the forecasts that analysts prepare and these, in turn, influence the share price. Some can even prepare models to estimate the likely impact that the messages given by their investor relations department will have on share price.

However, before we try to understand some of the tricks of the trade in successful investor relations, we first need to develop an appreciation of how the investment community works and how it measures shareholder value.

Shareholder value means different things to different investors. Although the basic techniques are all very similar, there are many differing interpretations of the term shareholder value. Companies need to appreciate this variety of interpretation if they want to communicate successfully with the investor community.

HOW DO THE FINANCIAL MARKETS WORK?

In Chapters 1 and 2 we explored the meaning of shareholder value. Now we will look at the way in which investment-relevant information flows around the financial markets. This will help us to identify *whose perception* you need to influence and *what information* buttons you need to press in order to protect and enhance your share price.

The financial markets are extremely complex and truly global. There are significant financial markets in the USA, Europe and the Far East and they each employ a variety of trading systems enabling shares to be bought and sold with ease. This improves the flow of business capital but also exposes shareholder value to increasing levels of scrutiny by sophisticated investors.

Shares can be traded on numerous exchanges around the world 24 hours a day. As a result, news that affects one stock market is likely to have a ripple effect on other time zones within a very short period of time.

Figure 3.1 shows a simple overview of the financial markets and the information flows that affect share prices. It highlights the key players in the investment community whose perceptions affect the shareholder value you are aiming to deliver. Information flows both *directly* from the company to fund managers and also *indirectly* through journalists, analysts and data providers – who together make up what we call here the investment 'information machine'.

Fig. 3.1 The flow of information in the financial markets

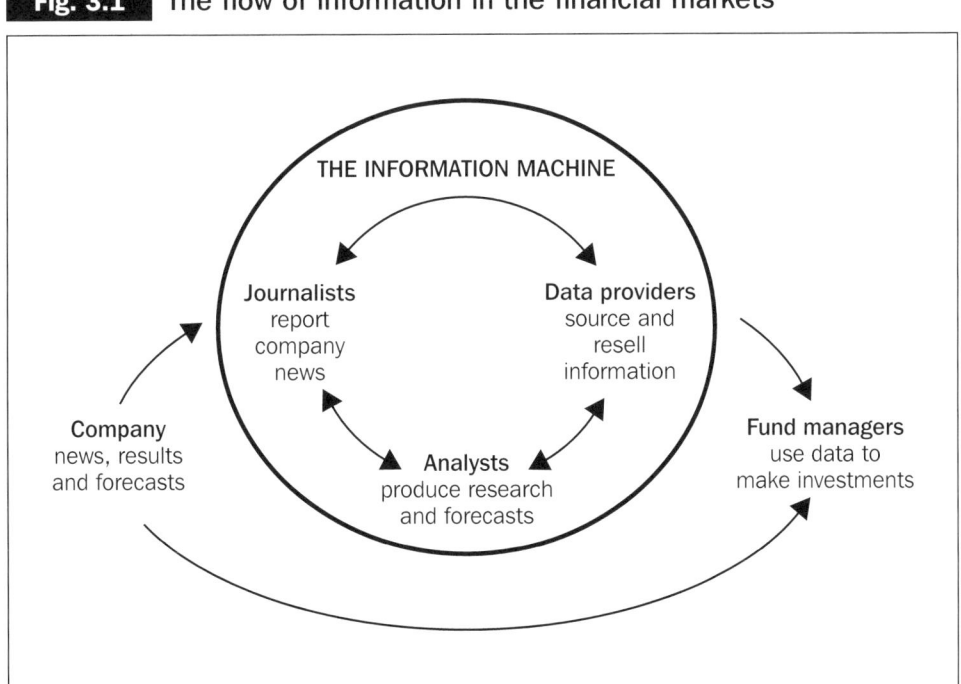

The main information flows within the markets are shown in Figure 3.1 and are as follows:

- The *company* provides press releases and briefings through its investor relations activities. It also releases annual historic financial results together with outlooks on future prospects.

- These are used by financial *journalists* and professional *analysts* who write about the company's recent and likely future performance. To complete their analyses they also use other information, such as economic forecasts and data about other companies within the same sector or industry. The analysts then generate their own forecasts of the company's likely earnings and cash flows over the next few years, normally in accordance with its business cycle. Analysts are found both within broker organisations (known as the 'sell side' because they sell their research) and fund management institutions, where they provide professional support to the fund managers.

■ Both sell-side analysts and journalists contribute their work to *data providers*, who collect textual and financial information from a variety of sources, in some cases adding value to it by calculating mean or 'consensus' earnings forecasts (the average expectations of all the analysts who track a particular company). They also provide other data items which enable, for example, comparisons of one company with an appropriate index, such as the FTSE 100.

■ The information flows between journalists, analysts and data providers are virtually circular. They sell and resell data, news and research to each other all the time, adding more value and updating their information on a regular basis. This way they oil the wheels of the 'information machine'.

■ The data providers resell their information, not only to other journalists and analysts, but also to the professional *fund management* community (also known as the 'buy-side').

■ The fund managers use the information they obtain both directly and indirectly to assist with their investment decisions, supported by in-house analytical reseach teams. Since fund managers are by far the major investors in stocks and shares they have a strong influence over the behaviour of the markets and the perceived value of any one particular stock.

While the *fund manager* is, in effect, the only investor shown in the picture, the roles of *journalists*, *analysts* and *data providers* also need to be examined carefully because they are important links in the information chain between the company and its shareholders.

WHO ARE THESE JOURNALISTS, ANALYSTS AND DATA PROVIDERS?

It is important for companies to know which journalists follow them and their industry. It is frequently the current and hot news about a particular company that has the most immediate impact on its share price. Financial journalists not only write in the business pages of major newspapers such as the *Financial Times*, they also contribute directly to some of the important news and data providers such as Dow Jones, Bloomberg and Reuters.

It is also extremely important for companies to develop good contacts with the analysts who research them and produce estimates of forecast performance. These estimates are the building blocks of the valuation of the company as perceived in the market. Analysts are found in a large number of investment institutions, such as Deutsche Bank, Paribas, ING Barings and Merrill Lynch.

In addition all companies need to develop a solid appreciation of the data provider community since they carry so much investment-relevant information about their business and its share price. They are an important source of data for the fund managers. The news providers above all sell directly to fund managers but there are also specialists who sell research and data. The two largest of these, First Call and I/B/E/S, merged in August 2000 to form the major player in the global market. Some other providers of estimates and/or broker research include Reuters, Thomson Financial Datastream, Multex, BARRA, Zacks and JCF.

THE NEED TO DEMONSTRATE SHAREHOLDER VALUE TO FUND MANAGERS

To influence investors, and be perceived as a creator of shareholder value, companies have to influence fund managers.

Fund managers are increasingly important players on the global investment stage. Over recent years they have become the major investors in the world and their dominance looks set to increase. Directly or indirectly fund managers control the investments of pension funds, investment trusts, unit trusts and insurance companies.

As professional investors, fund managers have to invest according to the requirements of their customers, those for whom they are managing the fund. In some cases this means that they will look for investments that generate a certain income. In others they may look for capital growth or a specific risk profile. Whatever their aim, it will affect their investment decisions, so it is not realistic to think that any particular company will necessarily be attractive to all fund managers.

Some funds are known as tracker funds. They adopt what is known as a passive investment style in that they seek to produce a portfolio that will mirror the movement in the stock market as a whole. Effectively this means that they need to maintain an investment portfolio that is a representative sample of the index as a whole. Therefore passive fund managers may seek to buy a company's stock simply to ensure that the volatility in its share price affects them in the same way as it does the whole market.

Active fund managers are a different breed. They believe in picking winners and selling losers. Using superior analytical techniques they aim to beat the market by their investment skill. Basically, they will buy a particular stock if they believe it will rise faster than others. So, if a company wants to 'sell' itself to these investors, it will need to demonstrate that it is in a position to deliver superior shareholder value.

Fund, or portfolio, management is of course more complex than this. Portfolio managers need to be aware of:

- *asset allocation* – the decision whether to invest in equities or other types of investment;

- *sector and country selection* – where to invest, i.e. in which countries and industries;

- *stock selection* – which shares to pick within the countries and sectors they have chosen;

- *risk management* – how to maximise the return for a given level of risk;

- *meeting performance targets and achieving preset portfolio benchmarks and investment constraints.*

All of the above issues affect the investment decisions of fund managers. As a result they also impact on the attractiveness of the shares in any particular company, and hence its share price. Companies need to be sure, for example, that when a fund manager examines a particular sector for a potential investment he will rate them highly in relation to their peers.

Comparisons with a particular sector are done through the use of peer groups and sector aggregates (the average performance of a number of comparator companies). However, most of the information that fund managers use for this purpose is forecast data, whether in numeric or text form. The reason for this is simple and relates to DCF principles: *forecasts drive share prices.*

Fund managers do not only rely on their own in-house analysts. They also have access to the 'information machine' described earlier. So it is clearly important for companies to put themselves in an influential position in relation to all these key players.

However, fund managers are too numerous to influence individually or non-selectively. Companies need to be clear about *who* they want to influence. They must also be clear about matching their message to the concerns of their potential investors. It is necessary to ensure that the company has a good story about its business and the economic environment within which it operates (customers, local markets, competitors, acquisition strategy, etc.).

We will now look in a little more detail at the systems and information used by fund managers and their in-house analysts. We will also highlight some of the key techniques that help measure shareholder value, in terms of target share prices.

THE IMPORTANCE OF BROKERS AND DATA PROVIDERS

The leading information providers distribute complex sets of data for use by fund managers. Most of the forecast information that they carry has at one stage been prepared by an analyst. So it can be argued that 'sell-side' analysts are key influencers in the investment information cycle.

But exactly what data items do analysts use to measure shareholder value? And what data items do information providers carry?

In the good old days life was simple. Companies declared their results and investors valued their shares based on their balance sheets or a multiple of profits. Then along came EPS and P/E ratios – so investors used those instead.

Today, life is very different. Some valuations are relative to a peer group and others are based on discounted cash flows. Analysts use a wide variety of measures in order to value companies. Sometimes it almost appears as though analysts are caught up in a continual rush to find new measures and insights that give one broker firm a competitive edge over another. Many of the measures that analysts use can also be found in one or more of the leading data providers.

Curiously, although many analysts are now adding cash-based valuations to their reports, most data providers still carry more information on forecast earnings than on forecast cash flows. Nevertheless, it is often possible to deduce a forecast cash flow figure from an earnings figure by making suitable adjustments for accruals accounting, working capital movements, depreciation and capital expenditure.

Many data providers now also carry a 'beta' rating for the companies they cover. The 'beta' rating measures the volatility, and therefore the risk, of the share in relation to movements in the market as a whole. It is one of the key variables in the calculation of cost of capital. In addition there is a wealth of information about local interest rates, government bond rates and market returns – all of these are also components of the calculation. So all the information is available to enable the calculation of cost of capital in the same way as that described in Appendix 2. All companies should make absolutely sure they know the returns that shareholders require them to deliver.

So, using the information carried by data providers, it is perfectly possible to perform share valuations based on discounted cash flows and thereby to measure the shareholder value any particular company is capable of creating. Therefore, in addition to influencing selected investors, companies should seek to press the buttons of those who distribute information to the entire investor community:

■ journalists;

- analysts – both in the broker (or 'sell' side) and within fund managers (or 'buy' side);
- data providers.

As part of the competitive race for capital, companies also need to watch:

- competitor performance;
- sector performance;
- analysts' output for them and their competitors;
- the data provided by information providers.

TWO IMPORTANT SCHOOLS OF SHARE VALUATION

Clearly, the investment community relies on a vast array of data and information systems and there is never one single measure that is commonly used by analysts and fund managers to value companies in a consistent way or to measure shareholder value. Instead a wide range of data is exploited to generate target share prices and buy/sell recommendations. This makes it particularly difficult for companies to manage the perceptions of potential and current investors.

Chapter 2 covered two different families of share valuation metrics: those that use *absolute* DCF methods and those that use non-DCF, or *relative*, methods.

In Appendix 5 we give two examples of analyst reports on the same company, Invensys PLC. The result of these two valuations cannot be compared because they were produced at different points in time. However, they are included to illustrate the two quite different applications of the separate families of share valuation metrics we described in Chapter 2:

- the first, by BNP Paribas, illustrates a *relative* valuation using traditional, non-DCF based measures;
- the second, by ING Barings, illustrates an *absolute* valuation using a DCF-based approach.

While these two valuation approaches are very different they are both valid. Both can reflect the expected future performance of the company. Analysts use each of them and frequently both methods can be found within the same valuation report. In such cases, the valuation one method produces helps to confirm the result produced by the other.

From the research reports in Appendix 5, it can be seen that companies need to manage their investor communications very carefully. Changes in anticipated EPS, profitability, EBITDA or cash flow will certainly affect the share valuations that analysts produce. However, the extent of that change, and the amount by which new information or revisions to forecasts will affect prices, depends upon the valuation methods being used by the analyst at the time. Since most public companies are followed by a number of analysts, each of whom may use differing valuation techniques, it should be apparent that companies need to understand:

- the specific valuation techniques their analysts are using;

- the likely impact on each of these valuation techniques when they release new results to the investor community.

USING INVESTOR RELATIONS TO HELP DELIVER SUPERIOR SHAREHOLDER VALUE

Companies should market themselves effectively to potential investors if they want to be perceived to be delivering superior shareholder value. Effective marketing, in the investor relations (IR) context, should incorporate the following:

- involving the CEO and CFO in a formal communications programme, including regular visits to major investment institutions – these will need to take place once or twice per year as a minimum;

- providing a contact point for important investors, using the IR department as a channel for access to the CEO;

- targeting those institutions that are likely to have a future interest in the company by:

  - making use of specialist surveys that are designed to highlight institutions which have a major position in the chosen sector;

  - searching specialist databases that highlight those institutions which have taken a position in a competitor within the same sector;

- evaluating whether the company would generate more shareholder interest if it were to change, for example, its dividend policy. There are specialist predictive models of fund manager behaviour that recognise the fact that some institutions prefer investments that deliver growth, value or dividends. By using these models companies can identify potential new investors;

- delivering a clear message about the future creation of value and the performance of the business by:

  - being prepared to argue why the company is relatively undervalued compared to its peers;

  - explaining that the company has adopted appropriate internal processes to manage for shareholder value or generate 'economic profit'. Fund managers do want to know that such processes have been implemented and that the company is genuinely managing for shareholder value. As yet, however, there still appears to be no need for *detailed* explanations as to how these processes actually work;

- actively 'selling' the company by focusing on future performance rather than historical results – do not be defensive just because the market may be volatile or troubled;

- avoiding surprises and communicating regularly – don't wait until the annual results are declared to deliver any bad news that may have a drastic impact on share price.

Although forecast results have a very strong influence on share price, companies would be well advised to appreciate the difficulties faced by the professional investor when using these estimates. Do not forget that some analysts are better than others!

Companies who focus some of their efforts on the better analysts are much more likely to reap the rewards of a more meaningful consensus within the investment community. Every year both Reuters and Extel publish surveys of analysts to identify those who are the most respected within the industry. So instead of a simple consensus forecast that represents the arithmetic mean of the estimates of all analysts, both good and bad, there is an emerging trend to select only those who are deemed to be the most accurate, thus creating an 'intelligent consensus'. Increasingly this is likely to be a major driver of share price.

There also appears to be an increasing trend for fund managers to rely on their own in-house analysts rather than on the estimates produced by others. This implies that companies who wish to manage their IR effectively should also seek to maintain good communications with the in-house analysts.

THE LESSONS THE INVESTMENT COMMUNITY TEACHES US

Successful investor relations is no longer about cosy lunches in the City. It's about developing a joined up approach to communicating the company's performance, marketing the business properly and systematically targeting potential investors. However, these activities also need to be built on the solid foundations that are covered in the later chapters of this book.

Companies who wish to demonstrate shareholder value must ensure they have a well staffed IR team, dedicated to managing the perceptions of investors and the information that flows around the investment information machine. They should, ideally, appoint someone within the IR team as the in-house 'expert' in shareholder value measurement and reflect this aspect in all investor communications. They should also consider using appropriate external IR expertise to help them target the most appropriate investors for the organisation.

However, it is important to tread a fine line in relation to the investors, information providers and analysts. While it is reasonable to review and correct information, companies must not unduly influence analysts or provide information to the benefit of one particular class of shareholder that may not be available to all equally. There is increasingly restrictive regulation in this area, particularly from the Securities and Exchange Commission (SEC) in the USA, and some companies are becoming nervous of falling foul of these regulations when they communicate with investors.

Companies should act proactively to 'manage' the impact of all news and earnings revisions upon the forecasts produced by analysts and, therefore, their share price. They should monitor the trends in the information metrics used by the analysts who follow their company and be prepared to discuss and comment upon them. This means:

- being prepared to adapt to the emerging trends in the information used by investors;
- staying ahead of the game and using the full variety of measures, especially earnings, cash flow (operations/funding/investment), EBITDA and truer, newer measures of shareholder value to model results;
- checking the analysts' forecasts and staying in touch with them, if necessary ensuring they have all the relevant facts at their disposal;
- checking that the data providers have correctly reproduced the analysts' forecasts;
- using specialist modelling tools and techniques to evaluate the impact of changes in results and dividends upon share price and upon the likely lists of potential investors in the company;
- calculating, before releasing results:
  - the earnings, EPS, EBITDA and free cash flow expected to be generated over the next five years;
  - the shareholder value the investment community is expected to perceive, using standard economic profit techniques, EV+ and TSR.

Ideally, companies should also be able demonstrate a clear link between the operation of their internal performance measurement frameworks and the creation of genuine shareholder value in terms of increased share price and dividends. The success with which companies can do this in practice will depend upon whether they have implemented a sound approach to VBM and managing for shareholder value. That is the third of our three perspectives.

However, before considering the design and implementation of the internal framework, we need to consider the second of our VBM perspectives, the use of value-based metrics to help evaluate strategy. That is the subject of the next chapter.

4

Using value-based metrics to evaluate strategy

CHAPTER SUMMARY

- When companies move to VBM they usually make changes at four levels – the corporate goal, strategic planning processes, business unit performance measurement and key performance indicators (KPIs).

- TSR is usually introduced as the corporate goal, over a period of years and relative to peer companies; this encourages management to focus on long-term cash generation rather than short-term indicators.

- Shareholder value thinking should encourage more quantitative approaches to strategic planning, using a TBR model to project the cash flow and evaluate the implications for adding value.

- These more quantitative approaches can help in comparing business values to share price, evaluating strategic alternatives and assessing the value created by long-term plans.

- A TBR framework will also allow the cascading down of targets, enabling delivery of strategy to be managed.

- Economic profit is a good overarching measure to give focus to that cascade because it encapsulates all the key value drivers.

- The use of economic profit can encourage the right strategies to be adopted. It encourages value creation, achieving returns which exceed the cost of capital.

- The cascade from TSR to economic profit can be extended further to incorporate value drivers and KPIs. Visual frameworks can be helpful in demonstrating and communicating this linkage within the business.

The indispensable first step to getting the things you want out of life is this: decide what you want.

Ben Stein

A FRAMEWORK FOR THE CREATION OF SHAREHOLDER VALUE

Companies can adopt a VBM approach in different ways and at different speeds. Those that have used the same advisers will, clearly, have some similarities of terminology and approach but, even with the same advisers, there will often be differences. For example Cargill and Unilever, two major international businesses, have both used the Boston Consulting Group for advice, yet have implemented quite different measures and approaches. Rightly so, because they are businesses with very different characteristics.

To examine the ways in which companies use VBM metrics as tools to evaluate strategy, we will first consider what needs to happen when a company adopts a value-based approach. This is shown in summary format in Figure 4.1, and it is represented as a 'cascading' sequence, from top to bottom. In practice, it may not always be implemented in this logical sequence.

Fig. 4.1 A framework for cascading value creation

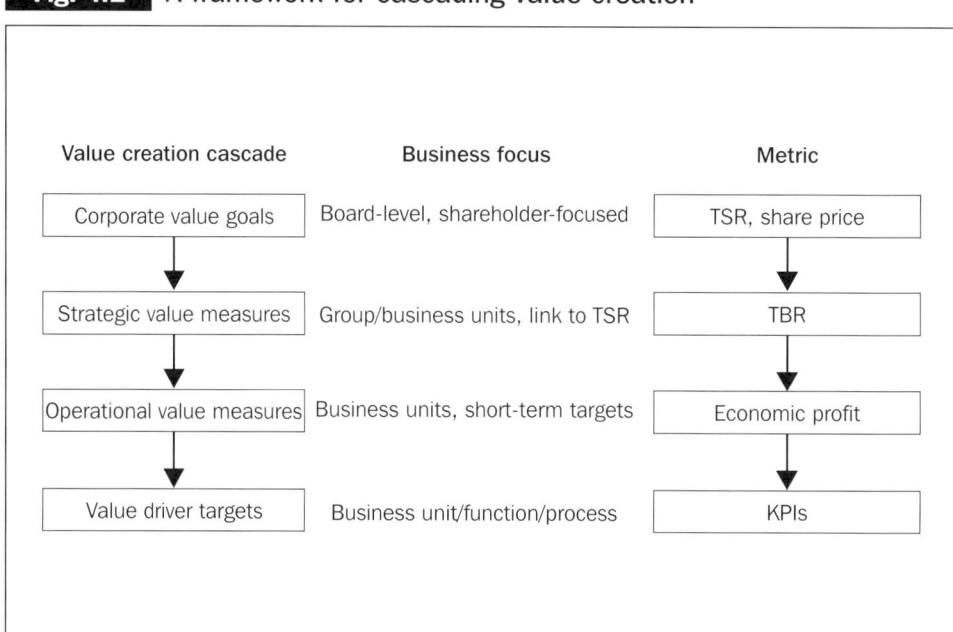

Value creation cascade	Business focus	Metric
Corporate value goals	Board-level, shareholder-focused	TSR, share price
Strategic value measures	Group/business units, link to TSR	TBR
Operational value measures	Business units, short-term targets	Economic profit
Value driver targets	Business unit/function/process	KPIs

The *corporate goal*, which should be agreed at the most senior level within the business, will normally relate to long-term shareholder value, usually based on TSR. If the company really means it, this goal will be linked to executive remuneration and will supersede the previous accounting-based measures. A company that claims to be value-based, and then declares that it is still paying management bonuses based on EPS alone, will not be seen as showing true commitment. Investors will also be more convinced of such claims if they are announced openly and publicly. For example, the announcement made by Unilever in October 1997, and referred to in Chapter 2, was important to both internal and external credibility.

At the next level of the cascade, *strategic value measurement*, TBR is sometimes confined to a few strategic planners. At other times it is more widely used within business units. Some companies use TBR at the outset, while others regard the use of TBR as a possible later stage implementation. This is because some companies have found that economic profit is a sufficiently powerful and flexible metric on its own and never feel the need to get into the greater complexities of TBR.

One feature of VBM is that it usually focuses on *short-term value measurement*. Often this results in the introduction of a value-based measure such as economic profit at the business unit or operating level. This stage is fundamental to changing the mindset and encouraging value thinking within business units. It is the key to making VBM real within the business and is often the first visible sign of a change at operating level. It can also be the first phase, even before TSR is adopted at top level. This may be desirable if there is healthy scepticism at the top and a prototyping, or piloting, approach is required to demonstrate the benefits of VBM and show how it will work in practice.

Even for companies who do not make the full link to TSR, the focus on maximising economic profit will encourage value-based thinking and provide a framework for value-based evaluation.

As mentioned above, the adoption of value-based approaches can have significant benefit if announced publicly to shareholders. Sony publicised their use of economic profit in April 1999. They also announced the adoption of 'Value Creation Management' as a driver of their strategy and an enabler of the creation of shareholder value within business units. These strategic changes, and the new way they intended to implement economic profit to measure business performance, were given real prominence at the beginning of their annual accounts report made by the president, Noboyuki Idei, and the chairman, Norio Ohga.

Finally, there should be a clear link from economic profit to a *value driver target* framework and a performance measurement cascade. This link will require a hierarchy of appropriate key performance indicators (KPIs), showing how decisions made by managers at the operating level will create value for shareholders. These KPIs should also become embedded within management objectives. An important tool, the 'balanced scorecard', is a key enabler of this cascading process and is defined in Chapter 5.

The following sections look at the impact that the introduction of VBM is likely to have at each of these four levels, highlighting the way in which this influences strategy evaluation.

HOW A CORPORATE TSR GOAL CAN HELP CHANGE THE PERCEPTION OF VALUE

The impact of the introduction of TSR as the key measure of corporate performance will very much depend on what has been in place before and what emphasis has previously been placed on share price and dividend payment. If the company's targeting has been too closely fixed to accounting measures like return on capital employed (ROCE) or EPS, the transformation can be fundamental. But in reality, it is unlikely that the top management of any major publicly quoted company could ever have failed to be motivated to increase the share price, even when EPS or ROCE were the main measures. Indeed, the need to grow EPS is likely to stem from the very desire to maintain or increase the share price, through achieving the 'double digit' growth that the market expects.

So how will the introduction of a TSR goal really change things? One obvious result may be an increased emphasis on dividend policies, particularly from companies that have previously adopted a policy of low dividend and high retained earnings. The questions that will now be asked are: 'Which strategy is likely to maximise TSR for shareholders? A high dividend? Or the retention of dividends for profitable reinvestment with the aim of achieving share price growth?' The answers will depend on investment opportunities and likely returns. Emphasis on TSR will encourage a focus on how to *maximise* value creation, rather than assuming that there is automatic virtue in low payout and high retention.

Perhaps the most significant impact of the introduction of TSR will be to encourage thinking about *relative* performance against competitors and alternative forms of investment. As with the BP case study in Chapter 1, the use of comparative TSR information will enable the directors to see themselves as the shareholders see them. Most corporate TSR goals are set in relative terms, against a basket of similar companies or against the market as a whole.

Knowing how you have performed against your peers can be a vital stimulus to encourage the setting of new targets and strategies. If a company is delivering lower TSR than its competitors because, for instance, the market analysts are not assuming future growth, this can concentrate the mind on strategies to achieve that growth.

The focus on TSR can also avoid the temptation to increase short-term earnings by cutting back on discretionary costs like research and advertising. This is an obvious way of increasing short-term EPS but the recent trend is for good stock market analysts to spot what is going on by detailed analysis of the accounts. Therefore, as the share price reflects their expectations of future cash flows, the impact on TSR is likely to be negative.

There may be a similar conflict around write-offs due to rationalisation and restructuring. These will naturally reduce EPS and this may be a serious barrier to taking the difficult decisions. Yet a true value-focused company will not give sole attention to the impact on EPS. It will also concentrate on communicating to the financial community the future benefits of the changes taking place. If this is done well, share price and therefore TSR, should rise.

The final impact of introducing TSR is that it provides the starting point for the cascade shown in Figure 4.1. The corporate TSR goal is the overarching framework for cascading long-term cash generation targets down into the business. To do this scientifically, and to embed it into the strategic planning process, we need the metrics of the next section.

USING TBR TO EVALUATE STRATEGIC PLANS

The key benefit of TBR as a strategic metric is that it enables clear quantification of the strategic options in present value terms and thereby provides management with a much clearer understanding of the factors that drive value. Assuming that TBR, or a similar value-based planning framework, is introduced, there are a number of ways in which it can be applied.

Indeed, companies develop a range of applications in order to evaluate their strategic options. Four of the applications for which TBR is typically used are as follows:

■ *Comparing the present value of future cash flows to share price.* TBR is commonly used as a metric to compare the stock market value of a company, or business unit, with the PV of its projected cash flows.

   This form of analysis can generate a variety of options. If the share price exceeds the PV value of future cash flows, top management can take the same route as Sony in 2000 and announce that the current share price is overvaluing the company. Such a direct message is unusual. The discovery of any such gap is likely to be followed by a more subtle managing of expectations. A more desirable route, if feasible, is to fill the gap by raising performance targets and then by delivering improved results. If the gap is the other way around, and the PV of future cash flows exceeds stock market value, then the challenge is to make sure that shareholders have the information they need to arrive at a similar valuation.

■ *Comparing present values of strategic options.* One option always worth considering is to evaluate how the PV of cash flows from the current strategy of a business unit compares to the option of disposal now, or at some time in the future.

   However, there are other options too. Comparisons could be made between new investments in fixed assets or the use of existing capacity, or between acquisitions and organic growth. TBR could also be used to evaluate the different impacts of growth and margin improvement. For example, the impacts on shareholder value of 2% volume growth and 1% margin improvement can be evaluated and compared. This can lead to better awareness of the potential for value creation and therefore better management decisions should result.

■ *Measuring the value of a five-year plan.* Most companies have some kind of long-term planning process which includes a total cash flow projection into the future. However, if there is no means of measuring the plan, it is impossible to say whether value will be created or destroyed by its implementation. A five-year plan which appears feasible could potentially be accepted despite the fact that it destroys shareholder value.

DCF-based models like TBR provide the mechanism and the formulae to calculate opening and closing values for the business. These evaluations, combined with the cash flows from the plan, make it possible to work out the present value created and the internal rate of return achieved. This enables the likely value creation to be assessed. If necessary the plan will have to be changed to ensure that it creates value in line with shareholder goals.

- *Providing the framework for economic profit and value driver targets.* Once a plan has been agreed, the essential task is to ensure its delivery. It is far too easy to forget strategic plans as managers work to operational targets that have no relation to them. An effective and integrated VBM system will ensure that the strategy is converted into annual targets, the delivery of which will ensure that the return in the plan is achieved. These become milestones, the passing of which provides assurance that all is on track.

   This application is the one that provides the vital link to the next stage in the cascade – that of measuring short-term performance through economic profit.

USING ECONOMIC PROFIT TO MEASURE THE VALUE BUSINESS UNITS CREATE

Economic profit can encapsulate many of the main value drivers in one measure. As a result, it is often the metric that is chosen by companies seeking to measure the value created by their businesses in any given period. This is because, although certain adjustments are necessary for a technically correct view of economic profit, it can be extracted relatively easily from financial and management accounting information.

Economic profit can also have a major role in influencing strategic priorities in many companies. The very fact that businesses are measured by economic profit has an impact on how strategic options are viewed. An illustration of this is provided below.

Figure 4.2 shows two divisions within a business, both measured by economic profit and both using a cost of capital of 10 per cent. Division A, with an ROCE of 30%, would appear to be very successful and management might not be motivated to seek ways to expand. In particular they might not want to become like B, which only has a 20% ROCE. However, if economic profit is introduced, Division B is shown to be more successful, creating 2½ times more value – £50 economic profit compared to £20.

The use of economic profit should motivate the managers of Division A to invest in projects which, although they may dilute ROCE, will still create increased economic profit and therefore add to shareholder value.

Fig. 4.2 Comparison of the economic profit of two divisions

	Division A £	Division B £
Results		
Post-tax profit	30	100
Capital employed	100	500
ROCE	30%	20%
Economic profit		
Post-tax profit	30	100
Less 10% charge on capital employed	(10)	(50)
Economic profit	20	50

Subject to risk assessment and normal project approval mechanisms, economic profit should encourage any manager to invest in projects for which the return exceeds the cost of capital.

The following example further illustrates this point. Figure 4.3 shows an existing business that is considering expansion.

Fig. 4.3 Using economic profit to evaluate expansion (1)

Results	Existing business £	New business £	Combined £
Post-tax profit	200	20	220
Capital employed	500	100	600
ROCE (profit/capital)	40%	20%	37%
Post-tax profit	200	20	220
10% charge on capital	(50)	(10)	(60)
Economic profit	150	10	160

If ROCE is used to assess the impact on performance, the opportunity would be rejected as it dilutes the company's ROCE from 40% to 37%. However, the economic profit number shows that the new investment adds value to the business.

Equally, when applied to a business with a low ROCE and a negative economic profit, the use of economic profit methods may prove that new opportunities, though increasing ROCE, will actually destroy value. This is illustrated in Figure 4.4.

In this example a company using ROCE would look favourably on the new business opportunity since the combined ROCE shows an improvement. The economic profit approach, however, shows that this decision would actually still destroy shareholder value. These two examples show that economic profit is a measure which supports growth, but only if that growth generates sufficient profit to create shareholder value by covering the cost of capital.

Fig. 4.4 Using economic profit to evaluate expansion (2)

Results	Existing business £	New business £	Combined business £
Post-tax profit	100	55	155
Capital employed	1,400	600	2,000
ROCE	7%	9%	8%
Post-tax profit	100	55	155
10% charge on capital	(140)	(60)	(200)
Economic profit/(loss)	(40)	(5)	(45)

There is one further use of economic profit for strategic decision-making. Many companies use economic profit as a proxy for long-term cash flow. As such, it becomes a relatively quick and easy way to measure the value creation implications of different strategic options, without going to the trouble and complexity of a cash flow model.

Further, products, customers and segments can also be evaluated in economic profit terms to assess their present and future potential for creating value. This can produce powerful insights. Cadbury Schweppes, for example, makes extensive use of economic profit presentation frameworks in the evaluation of their strategic options. They highlight where value is being created and how well they are performing relative to their competitors.

Such analyses are important contributions to strategic thinking. British Airways (BA) have used a similar economic profit measure, which they call Cash Value Added, to calculate the profitability of different business segments. For them, negative values are a clear signal of value destruction and they therefore seek to switch assets to the areas of its business which create positive value.

Although such analysis helps strategic thinking, it is not the ultimate decision-making mechanism. For example, BA still believes that the final decision to buy a new plane is still best evaluated through a full DCF-based project evaluation.

In summary, economic profit may support the evaluation of a range of strategic options but it should not necessarily be the sole decision-making tool.

SETTING TARGETS FOR THE DRIVERS OF VALUE

This is the final link in the VBM cascade. In some ways it is the simplest but it is also where it can all go wrong. This stage requires a sound understanding of the main drivers of value and their conversion into meaningful performance targets that can be used in practice by business units, functions and processes within the entire organisation.

These drivers can be described in many ways and can be converted into both financial and non-financial KPIs. Yet, apart from the cost of capital and the time value of cash flows, there are only five top-level value drivers that can emerge from a strategic plan. The simplicity of these five value drivers illustrates the paradox of VBM. It is highly complex at a *strategic* level, using metrics such as TBR within a detailed forecasting model. However, superficially it can appear 'blindingly obvious' when cascaded down to *operational* levels.

The five major value drivers at the operational level are shown in Figure 4.5.

Fig. 4.5 The five key value drivers

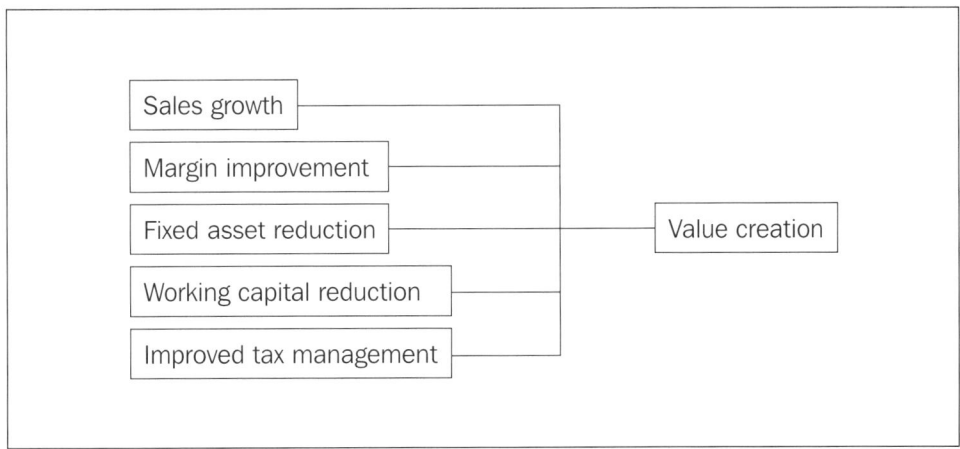

There is an obvious problem with such an apparently simple framework. When experienced managers are told that VBM distills down to these few value drivers, they will not find too much that is new. They may well point out that they already know that growing the business creates value and that improving margin by price increases or cost reduction is a good thing. They may accept that there should be greater emphasis on the control of fixed assets and working capital but that is unlikely to be a major revelation to high-calibre managers.

So the challenge of VBM is to overcome this barrier and convince managers at the sharp end that VBM is more than a recycling of obvious management principles. There are a number of ways in which this can be done and the success of this stage will determine whether VBM will really grab managers' imagination. The key is to prove to management that, despite its obvious nature, VBM really can be a new way of thinking and, perhaps, a new way of communicating what has to be done through the use of related management tools such as the balanced scorecard.

One powerful way to respond to the challenge is to introduce some visual frameworks which show the links between these value drivers in clear and meaningful form. Figure 4.6 is an example used in the VBM training manual at Boots. It is a simple but powerful model.

Fig. 4.6 Boots VBM value driver framework

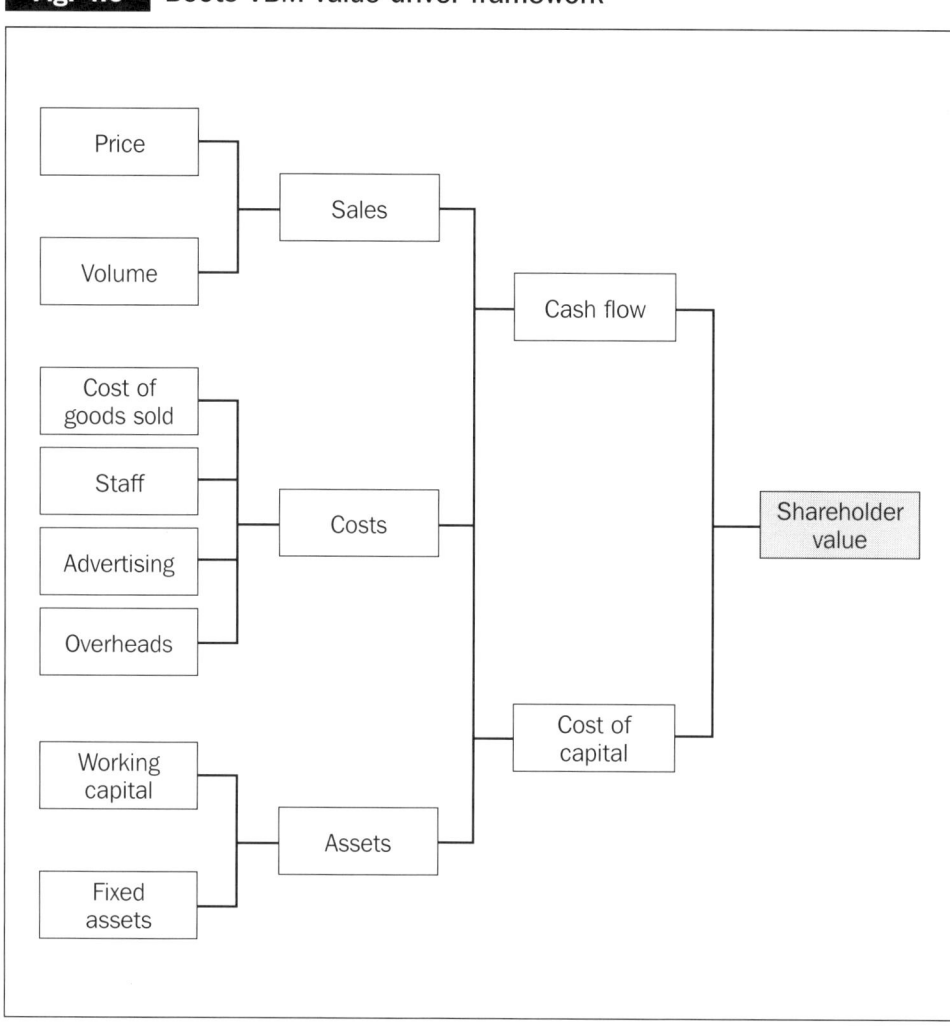

Frameworks such as these may not appear highly sophisticated but they are generally well received by managers at all levels. It is often the accounting staff who express the most reservations with such models. One group of financial managers in a major multinational criticised a similar framework as being 'just like the Dupont hierarchy of ratios'.

The managers were right about the framework. In many ways the model is similar to the Dupont hierarchy (see Figure 4.7), which is an alternative and equally powerful framework to communicate messages in terms of value creation. It helps to link together the drivers of value and to cascade value creation right down to operational level performance targets. Like the Boots model, it is always much appreciated by non-financial managers.

If VBM only provides the incentive to make financial managers demonstrate clearly the linkages between financial ratios and value creation it is still fulfilling an important role. For example, BP, whose process was highlighted in the case study in Chapter 1, combines ROCE with an earnings growth target. They believe this produces a close enough link to TSR to be used as the primary measure of value creation.

Fig. 4.7 Using Dupont to link value creation to operational levels

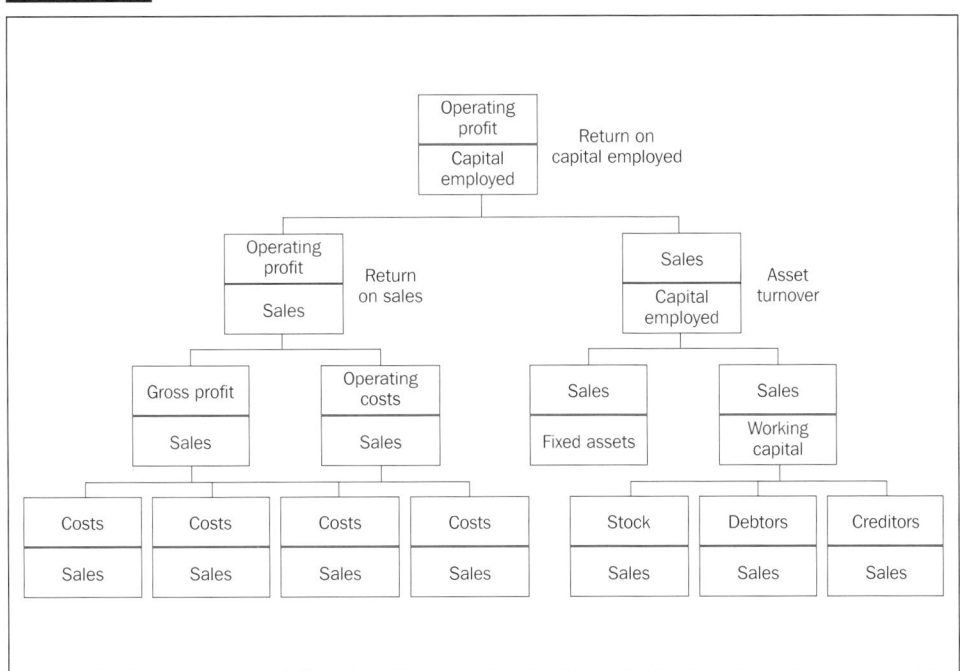

The other powerful role of such value driver frameworks is to show the link to *non-financial* measures, both those that directly drive the financial results and those that have an indirect longer-term impact on value. The framework shown in Figure 4.8 has been used in the business units of Unilever to present this wider measurement context.

It clearly incorporates a link from the value-based metrics, TSR and TBR, to the balanced scorecard framework, through the financial perspective of the scorecard. The balanced scorecard thereby enables the value creation concept to be communicated more clearly and cascaded throughout the business.

VBM has to consist of more than diagrams in order to be effective. The key benefits of its introduction come when such frameworks are developed further to confirm and display the linkages between the agreed KPI targets and the strategic goals. These cause : effect linkages are a key feature of the balanced scorecard and can be used to embed real understanding of the value drivers throughout the hierarchy. How much value does improving our KPI on stock levels create? How much do we need to increase margin to achieve our economic profit, TBR and TSR goals? How much value will we derive from an improvement in the rate of order fulfilment? These are the questions and answers that should be asked and answered in a VBM culture.

Analysis of this sort should result in clear decisions and action plans. It should enable the levels of achievement required to be embedded into the business, by establishing value-based objectives for functions, processes, teams and individuals. This should result in greater levels of confidence that, if everyone achieves their objectives, the benefits will flow from the drivers of value up to improved TSR.

Fig. 4.8 Unilever's framework for value-based management

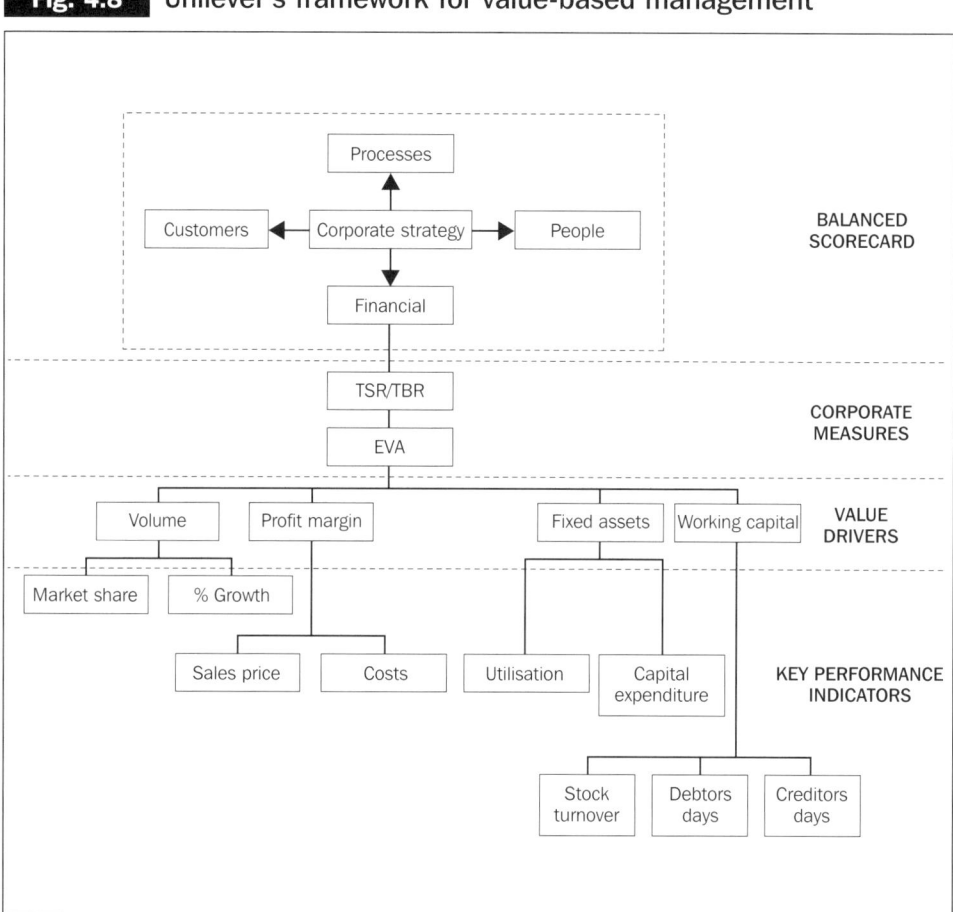

This is moving towards 'true' VBM: an integrated system that aligns the actions of everyone in the organisation to value and to shareholder goals. This is essentially the subject of the third perspective of the VBM framework, and the ways of achieving this are explored in the following chapters. They build on the two perspectives covered so far, 'managing the investor community' covered in Chapter 2, and 'evaluating strategies to create optimal value', the subject of this chapter.

5

Designing the measurement framework to deliver strategy

CHAPTER SUMMARY

- In designing the internal performance measurement framework to deliver strategy, the challenge is to clarify and interpret strategy, and understand how its value will be measured. Once this is achieved, a measurement cascade needs to be created and populated with data in order to focus communication below board level and inspire managers to deliver.

- Economic profit is often seen as the most appropriate metric for measuring value creation. As a result, it helps to focus the development of the internal measurement cascade.

- The balanced scorecard provides an equally important and pivotal link to strategy. It provides the framework for interpreting what has to be done to deliver the optimum shareholder value defined by the strategy. It is now also being increasingly used as the key enabling management tool that helps to create the measurement cascade.

- At a detailed level, activity-based costing (ABC) information can provide some of the necessary data for the balanced scorecard and management decision-making. It also provides important strategic insights, such as the relative profitability of products, services, customers, markets and channels. As a result, ABC can help managers to refine strategic decisions and model alternative 'what if' scenarios for improvement.

If you can imagine it,
You can achieve it.
If you can dream it,
You can become it.

William Arthur Ward

DESIGNING THE INTERNAL PERFORMANCE MEASUREMENT FRAMEWORK

Now is perhaps a useful point at which to paint a more complete picture of where we are within the three perspectives of our VBM framework – *managing the investor community, evaluating strategies to create optimal value* and *delivering value through IPM*. So far, we have covered two:

- The first was *managing the investor community* and, through Chapter 3, we discussed the issues and approaches in 'how to convey value to investors'.

■ The second was *evaluating strategies to create optimal value* and, through Chapter 4, we discussed how companies go about 'using value-based metrics to evaluate strategy'.

It is now time to turn our attention to the third perspective: *delivering value through IPM*. This comprises two key aspects:

■ *designing the measurement framework to deliver strategy* – the focus of this current chapter;

■ *embedding a VBM framework within the business* – this will be the focus of the next chapter and will essentially complete the VBM framework.

The book will then be completed with two final chapters, one focusing on the project management and change management challenge of implementing VBM, with the final chapter considering the future of VBM itself.

The aim of this chapter, therefore, is to outline the design of the internal performance measurement framework and cascade. The framework needs to be internally consistent between the corporate measures, the value drivers and the KPIs discussed so far. It also needs to focus on the measures that individuals or teams can be held accountable for in helping deliver shareholder value in line with the vision and strategy.

As discussed in Chapter 1, *this is where it could all go wrong with over 80% of companies failing to deliver on their strategic intent*. In designing the internal performance measurement framework to deliver strategy therefore, we need to ensure that it is capable of:

■ interpreting, clarifying and measuring the delivery of the value-based strategy;

■ enabling the production of a measurement cascade that helps focus energy on the communication of, and commitment to, strategy below board level;

■ being populated with quality data to provide the basis for real accountability and control.

This chapter now outlines some of the leading design principles and practices that meet these needs. They are based on many years of research, implementation experience, and indeed a little 'experimentation' in the field of advanced performance management, and they support the move towards adopting a leading practice approach. The principles and practices start from the premise that:

■ measures need to help drive real value creation rather than, say, just control costs or EPS growth;

■ performance drivers need to cross business processes and functions – this will make them more holistic with a stronger outward-facing perspective;

■ existing concepts and frameworks, such as ROCE, economic profit, the balanced scorecard and activity-based management, that may have been seen as stand-alone tools or solutions in the past should be integrated into one seamless management system over time.

Based on these principles, Figure 5.1 illustrates the concept of an integrated measurement and information framework and how it links to the third of the three VBM perspectives. It also highlights the key information components and their application. These information components need to be integrated over time to support VBM's internal business management approach for realising strategy. This conceptual framework is designed to help clarify how different types of quality information can help managers to realise strategy.

In simple practical terms, the information is clearly designed to support a number of management applications or processes around:

- making decisions;

- managing processes;

- measuring progress.

These applications are in turn dependent on:

- understanding the sources of 'true' profitability and where value is being created;

- planning the use of revenue and capital resources effectively;

- linking goals to activities so that it is very clear throughout the business unit what needs to be done and why;

- sharing information easily through effective reporting so that the right information gets to the right person at the right time.

Figure 5.1 illustrates how to design an information rich environment and thereby help realise strategy. It also gives a logical framework for defining how best to deploy and link a number of leading tools and techniques, some of which have been mentioned already. The following points highlight the dependancy on enabling techniques and methods:

- *Understanding the sources of 'true' profitability and where value is being created*. This is dependent on activity-based costing and shareholder/economic/ market valuation and modelling techniques.

- *Planning the use of revenue and capital resources effectively*. This is dependent on activity- and priority-based budgeting techniques, planning/modelling systems and project/initiative prioritisation techniques.

- *Linking goals to activities so that it is very clear throughout the business unit what needs to be done and why*. This is dependent on the balanced scorecard methodology.

Fig. 5.1 The integrated measurement and information framework

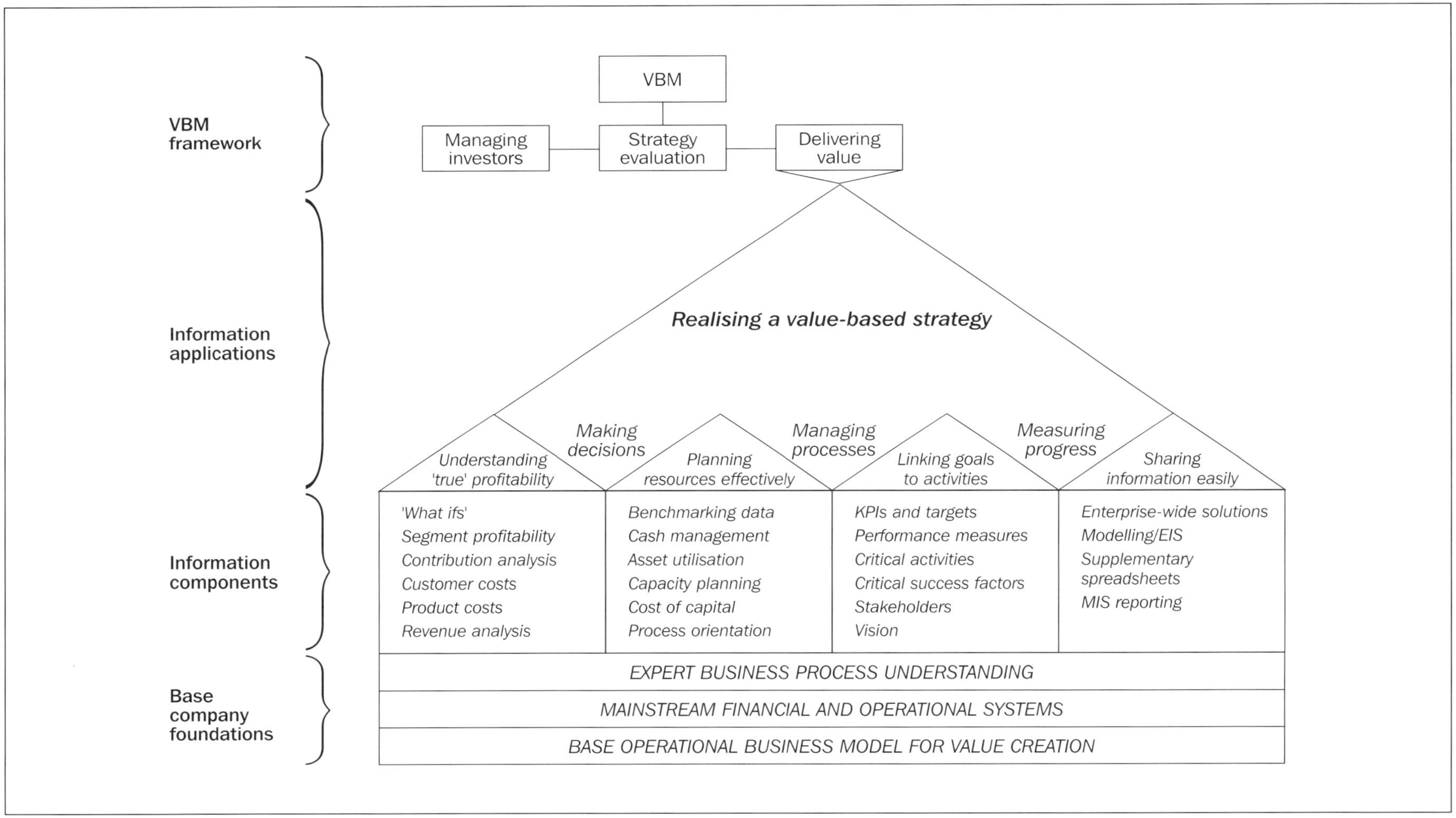

VBM framework

- VBM
- Managing investors
- Strategy evaluation
- Delivering value

Information applications

Realising a value-based strategy

Making decisions		Managing processes		Measuring progress	
Understanding 'true' profitability	Planning resources effectively		Linking goals to activities	Sharing information easily

Information components

Understanding 'true' profitability	Making decisions	Managing processes	Linking goals to activities	Measuring progress
'What ifs'	Benchmarking data		KPIs and targets	Enterprise-wide solutions
Segment profitability	Cash management		Performance measures	Modelling/EIS
Contribution analysis	Asset utilisation		Critical activities	Supplementary spreadsheets
Customer costs	Capacity planning		Critical success factors	MIS reporting
Product costs	Cost of capital		Stakeholders	
Revenue analysis	Process orientation		Vision	

EXPERT BUSINESS PROCESS UNDERSTANDING

MAINSTREAM FINANCIAL AND OPERATIONAL SYSTEMS

Base company foundations

BASE OPERATIONAL BUSINESS MODEL FOR VALUE CREATION

■ *Sharing information easily through effective reporting so that the right information gets to the right person at the right time.* This is dependent on business intelligence and decision support systems.

As can be seen from the above, businesses require several interdependent tools and processes to help deliver the above aims. In addition, various operational and financial systems are needed to support the core transactional activity of the business. This includes, for example, paying bills, processing orders, managing the supply chain, producing invoices, maintaining ledgers, etc. Without this foundation, it clearly becomes difficult to contemplate the more advanced information requirements. Further, process analysis, activity mapping and root cause analysis techniques are critical for building up the business process understanding that is so vital for high-quality information definition. We will further clarify and re-emphasise these points later.

While Figure 5.1 shows a high-level conceptual view of what the measurement and information framework needs to deliver, the detail is missing. It is now appropriate to examine the most important of the measurement tools and techniques that support this conceptual measurement and information framework as a 'world-class' information management approach. They include:

■ *economic profit* to measure shareholder value creation and help compare strategic options;

■ *the balanced scorecard* to interpret, communicate and cascade the chosen value-based strategy and help deliver the appropriate amount of shareholder value consistent with it;

■ *activity-based costing* to meet some of the more detailed costing, profitability and process-based information requirements.

These tools must be integrated to provide a robust management information framework. Guidelines on how to do this are given in the following paragraphs. However, the pace and degree to which these are fully adopted will reflect the business needs of each individual company's circumstances and business priorities. As a result, experience suggests that the framework will be applied in a bespoke way due to differing priorities in each unique set of circumstances.

ECONOMIC PROFIT AS THE OVERARCHING INTERNAL METRIC

As discussed in earlier chapters, shareholder value measurement techniques can be based both on internally focused financial measures and external or market-based measures. Market-based measures are the real test of whether the ultimate aim of delivering shareholder value has been or will be achieved. In contrast, internal measures can provide management with an ongoing guidance system to support decision-making, and planning and control in an integrated way.

Economic profit is one such an internal measure and it has been widely adopted as part of a typical VBM implementation approach. To illustrate the measure, Figure 5.2 shows the 'economic profit tree' and the clear relationship structure between the various financial data items within the structure of the tree.

Because economic profit is based on accounting numbers, it is possible to implement this aspect of a value-based financial management system quickly as a simple extension to typical reported outputs. The only absolutely essential minimum adjustment needed in a conventional profit and loss account is the deduction of an appropriate capital charge.

Managers already familiar with the notion of profitability are likely to be able to adjust relatively quickly to this revised approach. In addition, the costs and time involved in adapting the accounting system to incorporate this adjustment are likely to be insignificant in most cases.

Economic profit is the first internal building block of the measurement and information framework. It shows how value-based metrics can start to be broken down into more detail to support the principles and requirements of VBM, including, for example, the value driver analysis mentioned in the previous chapter. As such, it is the basis for an internally consistent cascade of linked measures.

THE VALUE OF THE PROCESS PERSPECTIVE

Figure 5.1 highlighted the need for a good appreciation of business processes as being the key to building up an excellent measurement and information framework. So, how do we create this good appreciation? Figure 5.3 shows the high-level process model that was used to cascade processes and activities at a major telecommunications company. This approach was used to help develop an integrated performance measurement and management system within that company.

However, an integrated performance measurement and management framework clearly needs to cover the processes of the whole business. Figure 5.4 illustrates another process cascade. It shows the complete process framework that was used within a typical food processing business.

Now that we understand the economic profit tree and a basic process model of the business, it is time to consider how the key strategies of the business can be measured, cascaded and communicated throughout the business to meet the strategic value-based targets. This is achieved through the technique called the balanced scorecard.

Fig. 5.2 The economic profit tree

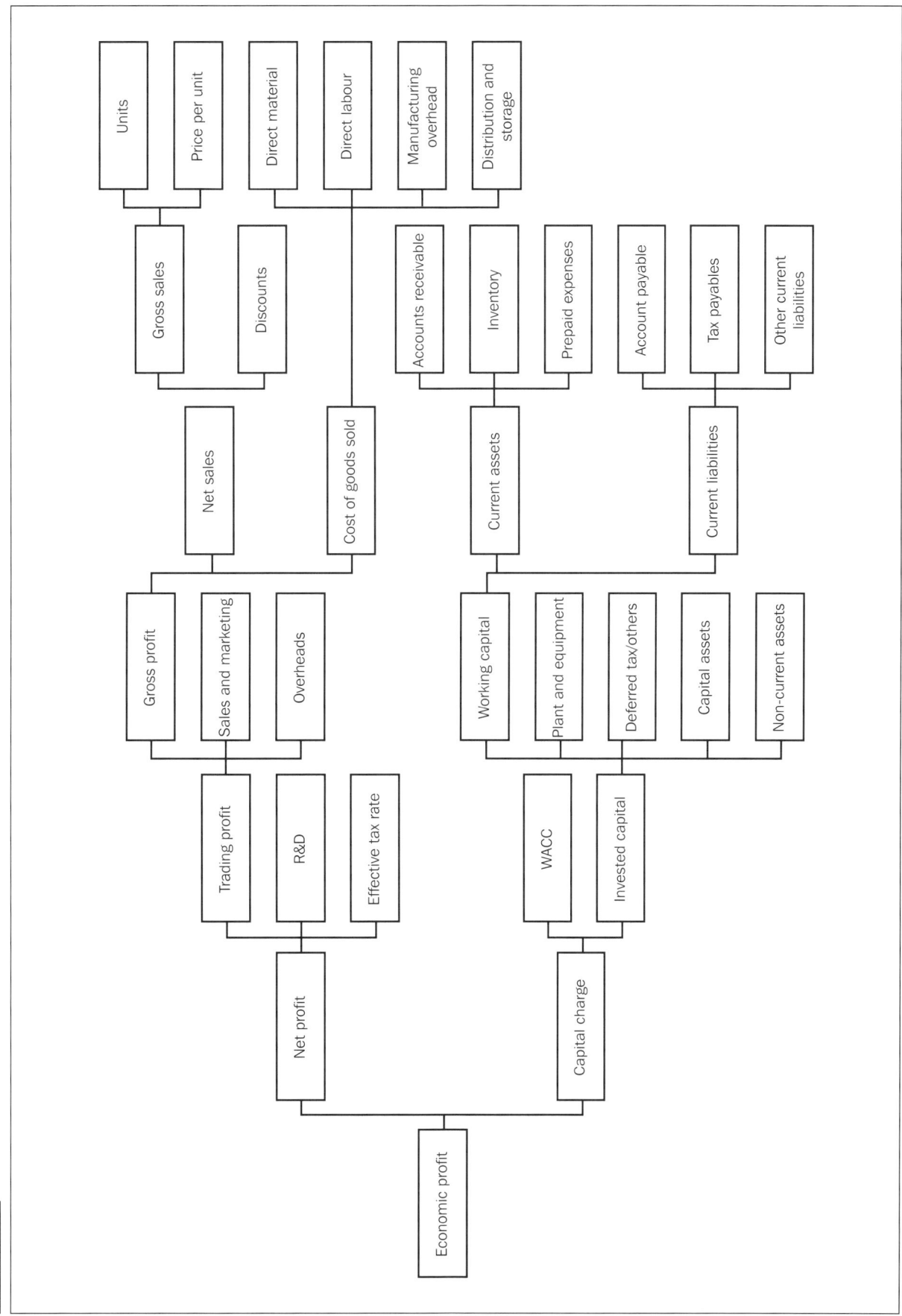

Fig. 5.3 Understanding the process perspective

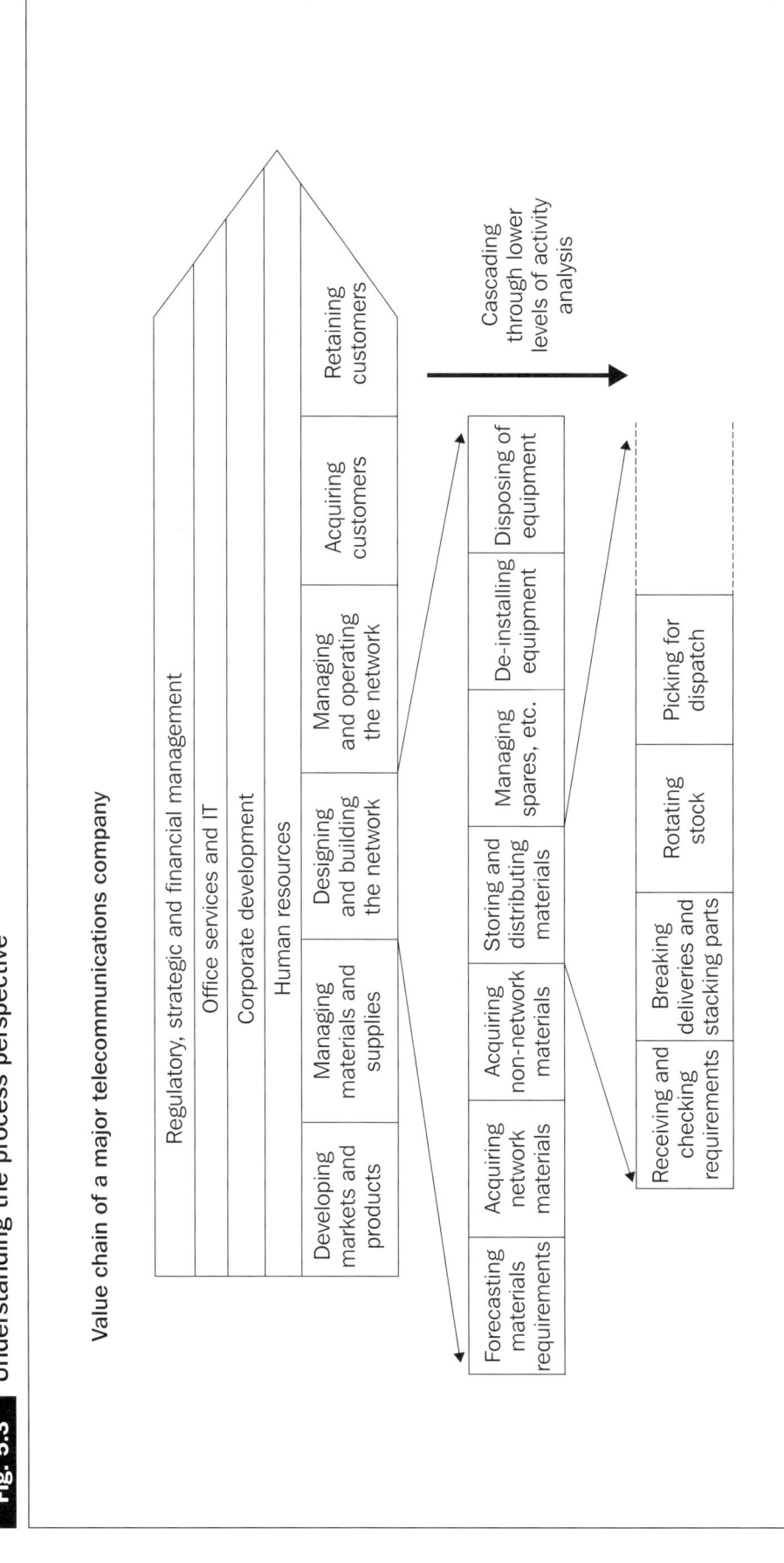

Value chain of a major telecommunications company

Source: Ashworth, G. and Gwynne, R. (1993) 'Implementing activity-based management,' *Management Accounting*, December

Fig. 5.4 Completing the process perspective

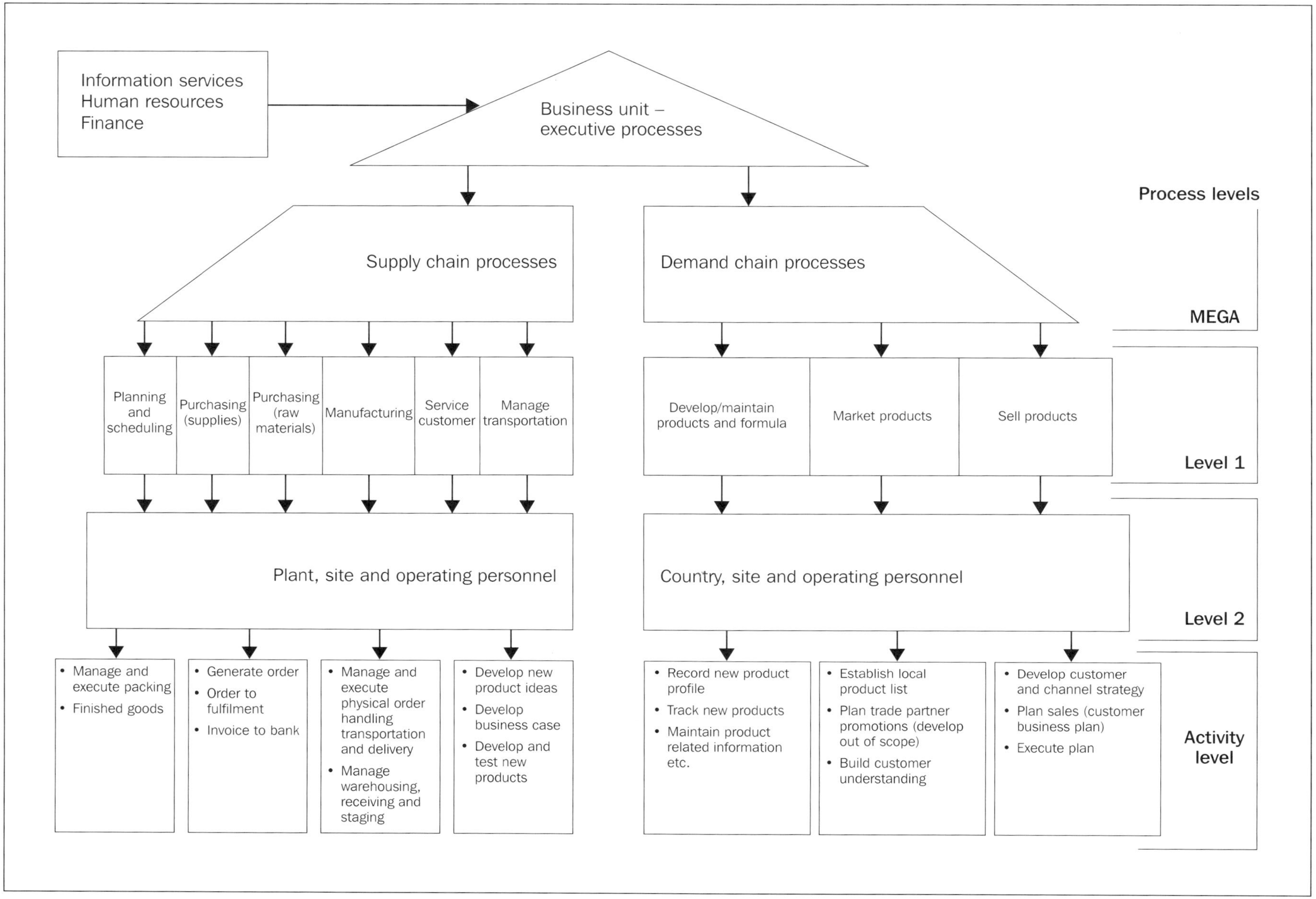

THE PIVOTAL ROLE OF THE BALANCED SCORECARD

The balanced scorecard has been around now as a management tool for the best part of a decade and its acceptance and use in major companies is already quite extensive and still on the increase.

In essence, it provides a framework that can help to interpret strategy through four interlinked perspectives that are often referred to as the 'value creation imperatives'. These four perspectives are:

- financial;
- customer;
- internal business process;
- people, innovation, improvement, learning and growth.

Sitting above the balanced scorecard perspectives must be the vision and strategy of the business. The vision sets the longer-term direction of where the company aspires to be and on what basis it will compete, and also sets the 'blueprint' for strategy formulation and evaluation, as discussed in Chapter 4. The value-based strategy that emerges must then be interpreted within each of the four perspectives or 'value creation imperatives' of a business, so that there is:

- a set of 'critical success factors' or CSFs, which answer the question, what are the critical things that have to happen so that the strategic objectives can be achieved successfully?
- a set of related measures that help to monitor how well the CSFs are being achieved. The measures need to focus not only on past performance, but also future performance, often referred to as lagging and leading indicators respectively. Typically, there may be around 20 to 25 such measures, at most, at the strategic level to help communicate the strategy across the four perspectives;
- targets for each of the measures;
- related initiatives to help ensure realisation of the targets.

This results in a comprehensive framework that helps to translate a company's vision and strategy through a coherent set of linked objectives or CSFs and performance measures. Some companies also then seek further insights by linking measures to a small number of strategic themes for the organisation, for example grow the business, or increase productivity.

In short, measures within the balanced scorecard help to articulate business strategy, and they help to align individuals with organisational and cross-departmental initiatives to achieve a common goal for the future. This is one reason why a good appreciation of business processes is important. Further, one of the four perspectives of the balanced scorecard is focused entirely on the 'internal business process' and this is clearly an important means by which a subsequent performance measurement cascade can be achieved.

However, at the strategic level, the four perspectives help achieve a balance between:

■ short and long-term objectives;

■ desired outcomes (often most associated with the financial and customer perspectives) and the performance drivers of those outcomes (often most associated with the internal business process and people, innovation, learning and growth perspectives). This thinking helps build a better understanding of causes (drivers) and effects (outcomes);

■ hard (or quantitative) and soft (or qualitative) measures.

In order to demonstrate the role of the balanced scorecard in contributing to a process-based measurement cascade, and in creating cause and effect understanding, we will now look at a simplified example from a financial services company (FSC).

CASE STUDY – BALANCED SCORECARD

FSC has a radical new vision. It aims to turn round the business through a strategy based on generating significant growth. It aims to get:

■ significantly more customers;

■ more products to the new and existing customers through cross-selling.

A key strategy is to use its call-centre as a means of building superior customer relationships through the quality of the customer experience, enabled by a sophisticated customer management database. This represents a major shift of emphasis for the call-centre from dealing with problems and complaints to managing relationships for developing business.

It aims to achieve its new vision through 12 key statements of strategic intent or critical success factors:

■ Increase value of members' fund well in excess of the all-share index.

■ Achieve lowest cost ratio within the sector.

■ Treble sales in the next four years.

■ Double customer portfolio in three years.

■ Ensure that all regulatory and compliance requirements are met.

■ Meet stringent customer satisfaction criteria.

■ Meet customers' needs effectively.

■ Provide leading product portfolio.

- Cross-sell as a basis of growth and through quality of relationship.
- Increase staff skills to meet the challenge.
- Invest in technology to enable and support customer–staff interaction.
- Seek out new ways to ensure maximum investment returns.

Having developed these critical success factors, the next step is to identify how to deliver them. To do this, FSC ran a series of interviews and workshop sessions. During these sessions, the following questions were considered by the management team:

- What are our core capabilities deemed to be to deliver our strategy? (*internal business process perspective*)
- How are we going to improve these core capabilities to achieve real and sustainable competitive advantage? (*improvement and learning perspective*)
- What impact are we trying to have in the market, witnessed by the measurable impact on our customers? (*customer perspective*)
- How will this be reflected in the financial results we want to achieve and the economic profit we are seeking to create? (*financial perspective*)
- What are the relationships and dependencies we expect to see between each of these things? (*overall business model flow*)

The management team's goal in considering these questions was to create a model of the business strategy that was fully agreed among them. The model would give them a clear picture of what the business was trying to achieve and how the various statements of strategic intent linked together and supported the overarching value creation goal of 'increasing the value of the members' fund well in excess of the all-share index'.

The business model resulting from the interviews and workshop sessions is shown in Figure 5.5. It could be viewed as a flow diagram that fully reflects the 12 key statements of strategic intent. The benefit of this view, however, is in the clarity of the cause–effect relationships that have been established between what needs to be done internally in order to make an impact externally.

Once the clarity of the business model has been established, a set of leading and lagging strategic measures can be defined and agreed. This is the next step in the process of defining a balanced set of measures that together comprise the balanced scorecard. A balanced set of measures should help, not just to assess how well you have done, but how well you can expect to do. This is particularly true if your judgement on the cause–effect relationships and dynamics within the business model itself is sound.

For example, if you are on track with your planned investments in innovative approaches to beat the all-share index, then ultimately you would expect it to be reflected in your actual performance relative to the index. A more complex example illustrates the point further. If you improve your forecasting accuracy on demand levels in relation to resourcing levels for 'Dealing with members efficiently and effectively' within the call centre, then you should see this as a contributory factor in ultimately achieving the lowest cost ratio in the industry. In addition, and at the same time, you should also see the service quality being maintained and reflected in the following measures:

■ customer response times less than 15 seconds;

■ customer forum feedback;

■ customer satisfaction surveys;

■ customer churn ratios by month.

Fig. 5.5 Strategy interpretation – the business model of FSC

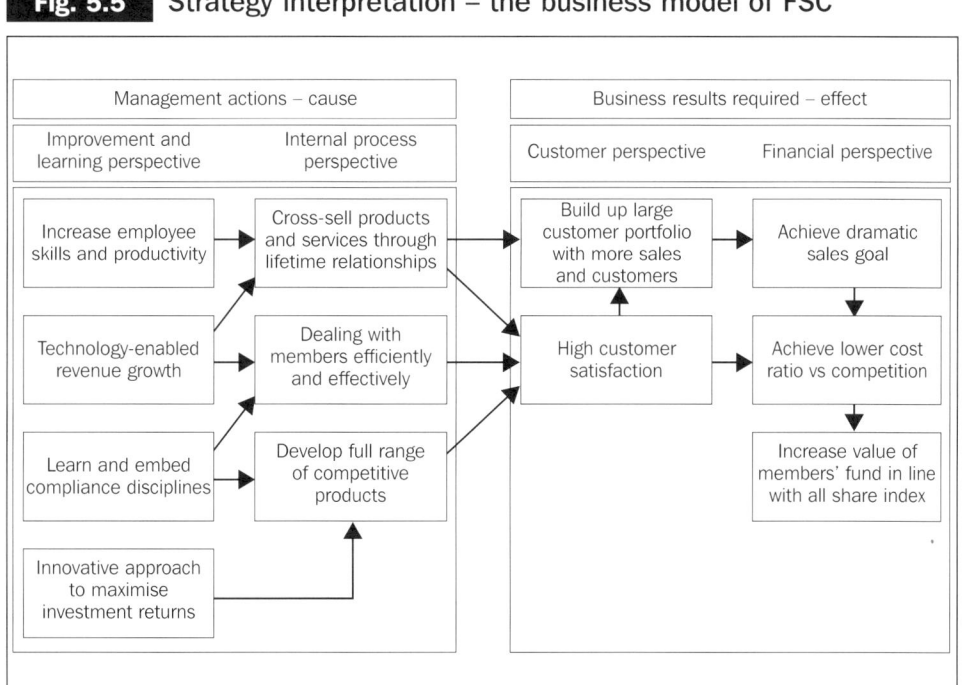

This is because resources are being deployed exactly as required due to the quality of the demand forecasts. However, if progress against these linked measures is not being achieved, you may have already started to identify why and to determine what you need to do about it. The full range of leading and lagging measures that FSC created are shown in Figure 5.6.

Strategic measures are extremely valuable in helping executive teams to assess the health and future prospects of the business. In themselves, however, they are not enough. Why? Too often things do not always go according to plan and corrective action needs to be taken. It is vital therefore to have a measurement cascade that gives an ability to focus on what needs to improve in order for you to get the results required. This has started to be explored at the strategic level, but for robust and effective business management, we need to go a little further.

Fig. 5.6 Creating a balanced set of measures

Strategic intent	Lag indicator	Lead indicator
Increase value of members' fund in line with all share index	Relative performance vs growth trends	Investment in innovation
Achieve a lower cost ratio vs competition	Cost trend by month to target by sector	Stretch target vs benchmark
Achieve dramatic sales growth	Growth rate achieved vs that required by week to hit target	% of new sales from existing customers
Build up large customer portfolio with more sales and customers	Growth rate achieved vs growth rate required by month to hit 5m members by year 2001	Marketing campaign programme variance
High customer satisfaction	Satisfaction surveys. Customer churn ratio by month.	Customer forum feedback
Develop full range of competitive products	% sales from new products. No. of new products	Average time to market of new products
Dealing with members efficiently and effectively	Responses less than 15 sec vs target	Forecasting accuracy on demand levels
Cross-sell products and sources through lifetime relationships	Average number of products held in key customer target segments per month	Price basket vs customer value index
Innovative approach to maximise investment returns	Investment trend	Size and quality of investment team (competency index)
Learn and embed compliance disciplines	Trend on reported compliance issues	No. of training days accredited vs target
Technology-enabled revenue growth	Key projects delivered to time, on budget and realising promised benefit	Value of business cases approved
Increase employee skills and productivity	Staff survey results	Competency index. No. of training days completed vs target

Figures 5.7 and 5.8 show one way that 'cascading' can work in practice. Figure 5.7 illustrates how the measurement cascade can be built below the executive team. It shows the underlying measures that contribute to the leading and lagging strategic measures for one of the business model components. This is the strategic intent or CSF named: 'Cross-sell products and services through lifetime relationships'. Through understanding what the subordinate key strategies are for meeting this CSF, it is possible to define the measures that relate to them. It is also then possible to identify where the responsibility lies for each of these measures.

Fig. 5.7 Building the measurement cascade

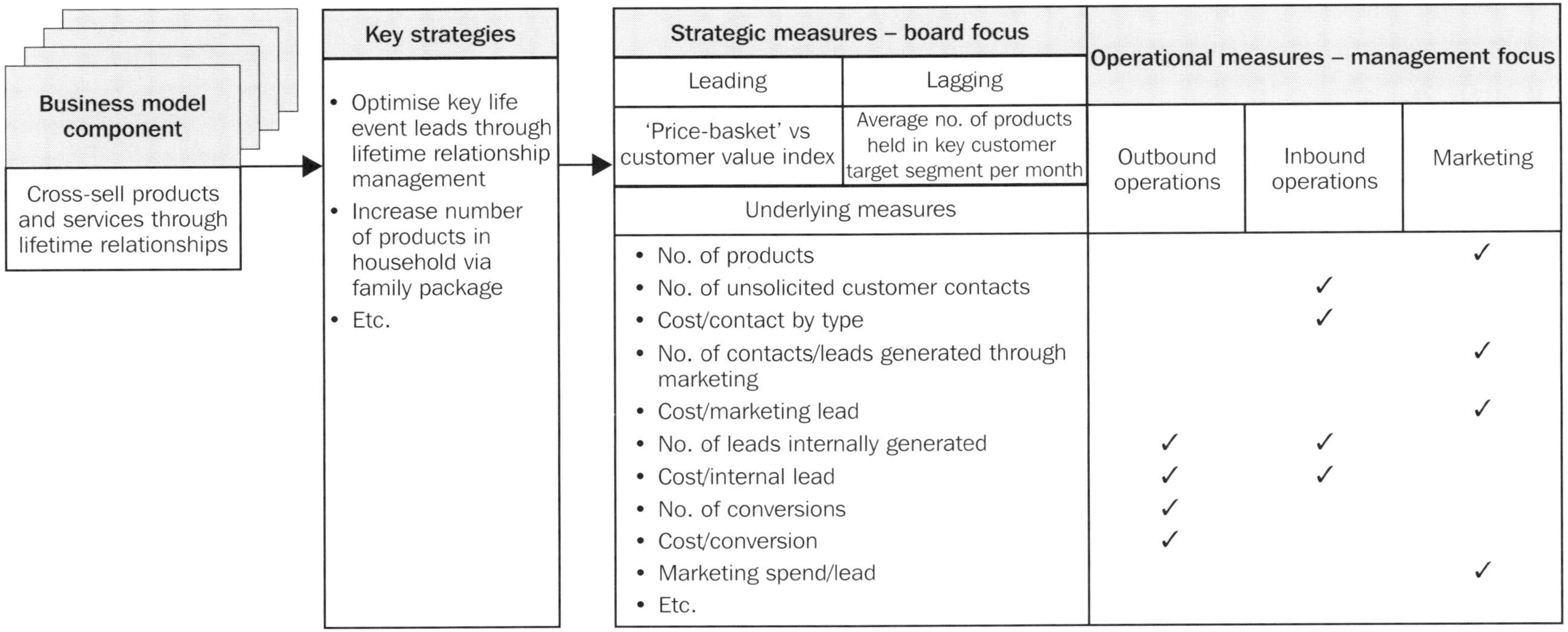

Business model component	Key strategies	Strategic measures – board focus		Operational measures – management focus		
		Leading	**Lagging**	Outbound operations	Inbound operations	Marketing
Cross-sell products and services through lifetime relationships	• Optimise key life event leads through lifetime relationship management • Increase number of products in household via family package • Etc.	'Price-basket' vs customer value index	Average no. of products held in key customer target segment per month			
		Underlying measures				
		• No. of products				✓
		• No. of unsolicited customer contacts			✓	
		• Cost/contact by type			✓	✓
		• No. of contacts/leads generated through marketing				✓
		• Cost/marketing lead		✓		
		• No. of leads internally generated		✓		
		• Cost/internal lead		✓	✓	
		• No. of conversions			✓	
		• Cost/conversion				
		• Marketing spend/lead				✓
		• Etc.				

Fig. 5.8 Establishing the process/team/individual responsibility matrix

Process description	Process owner	Performance measures and improvement planning									
		Financial	Target	Current	Non-financial	Target	Current	Root cause	Action	By	Progress update
Marketing	P. Pratt	Reduce fixed marketing cost per £100 of revenue	£10.00	£14.74	Increase no. of products	100	75	Product development process failing to meet plan	Adopt and roll out best practice product development process	AB	30/4/2002
		Reduce variable marketing cost per lead	£25.00	£28.70	Increase no. of leads generated	1000	723	Campaign response effectiveness	Evaluate alternative campaign tactics and follow-up procedures	BA	30/6/2002
		Increase sales growth per annum	40%	25%							

For example, the 'Number of products' is deemed to be one measure that contributes to achieving progress on a higher level metric of 'Average number of products held in key customer target segment per month'. This is then shown to be the responsibility of the marketing process owner, along with a number of other marketing-based measures.

Following the cascade trial, Figure 5.8 now shows an illustrative extract of the marketing responsibility matrix that then needs to be used within this area of operation. The matrix also shows how the measures can be used to drive improvement. For example, it is clear that the number of products are well down on target by 25%. This is also, perhaps, contributing to a number of the other performance measures lagging behind target and emphasises the need for action to be taken – adopting and rolling out a new or best-practice product development process. This clarity also then helps form the basis for real accountability and stewardship.

HOW ACTIVITY-BASED COSTING CAN HELP

Not all of the measures illustrated in Figures 5.7 and 5.8 would typically be available from the core financial reporting systems, for example cost per contact by type, cost per conversion, core process cost %, etc. So, how are they generated?

To calculate some of them, data may be required from operational systems, external sources or surveys. However, from a costing and profitability perspective, this is where the 'ABC' helps. It illustrates why an inextricable link may need to be built between economic profit, balanced scorecard and the performance measurement cascade, and ABC.

ABC enables us to:

- understand activity and process costs and what the key cost drivers are behind them;
- produce meaningful product, customer, service, channel cost and profitability information;
- identify opportunities for eliminating non-value added or diversionary costs;
- support an ongoing framework for resource and performance measurement and management.

It is dependent upon carrying out a comprehensive activity analysis within a process model of the business as discussed earlier. This is key to constructing the relevant activity cost information and the activity focus because:

- it is the lowest common denominator between a function and a process;
- it describes what the organisation does and what it costs to do it;
- it provides the basis for measuring the cost of its outputs.

A detailed review of ABC is beyond the scope of this book. However, to help complete the picture of the role of ABC, Figure 5.9 illustrates the typical information processing aspects and outputs of an ABC calculation model. The ABC model captures the key business processes at an activity level and this is often defined within an 'activity dictionary'. It also takes cost data from the core financial systems. Then, based on an understanding of how resources are used in the business, it allocates costs to activities and the cost of activities to products, services and customers based on activity cost drivers. It therefore becomes a rich source of costing data to support the required information applications effectively and gives unique insights to the many and complex sets of activities performed within the business.

Fig. 5.9 A model for calculating ABC information

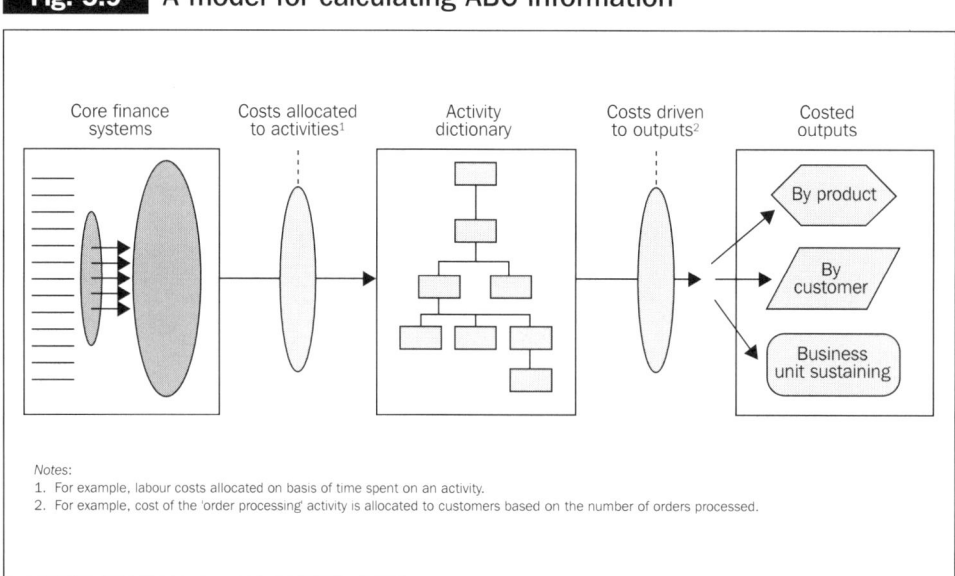

Notes:
1. For example, labour costs allocated on basis of time spent on an activity.
2. For example, cost of the 'order processing' activity is allocated to customers based on the number of orders processed.

PULLING IT ALL TOGETHER

The important issue is to link ABC with the balanced scorecard and economic profit in the most appropriate way for the unique circumstances of each business. This needs to be done so that value and the drivers of value can be consistently and comprehensively understood and measured over time.

Within the latter part of this chapter, we have tried to highlight how ABC is needed to help populate some of the measurement framework cascaded within the balanced scorecard. But how does this link back to the economic profit tree? Figure 5.10 now shows an overlay of how this works for a major food manufacturing company. It shows the standard economic profit tree and how a small sample of the ABC and balanced scorecard metrics can be easily related to the key financial measures within the 'tree'.

It is now possible to create an internal performance measurement framework and cascade that is consistent with a range of financially orientated, value-based metrics. Clearly, none of the three key measurement techniques discussed in this chapter can be used as effectively in isolation as they can in combination. They all complement each other to create an integrated measurement and information framework. This chapter has given an overview of, and some guidance on, how this can be achieved and what it looks like.

A key phrase in the above paragraph is 'can be used'. To get value from this guideline performance measurement framework, it must be used properly to drive benefits. This means that it must be designed specifically for each unique business. The challenge of then embedding and aligning such a framework, to achieve lasting and sustainable benefits within a business, is the subject of the next chapter.

Fig. 5.10 Overlaying the measurement cascade onto the economic profit tree

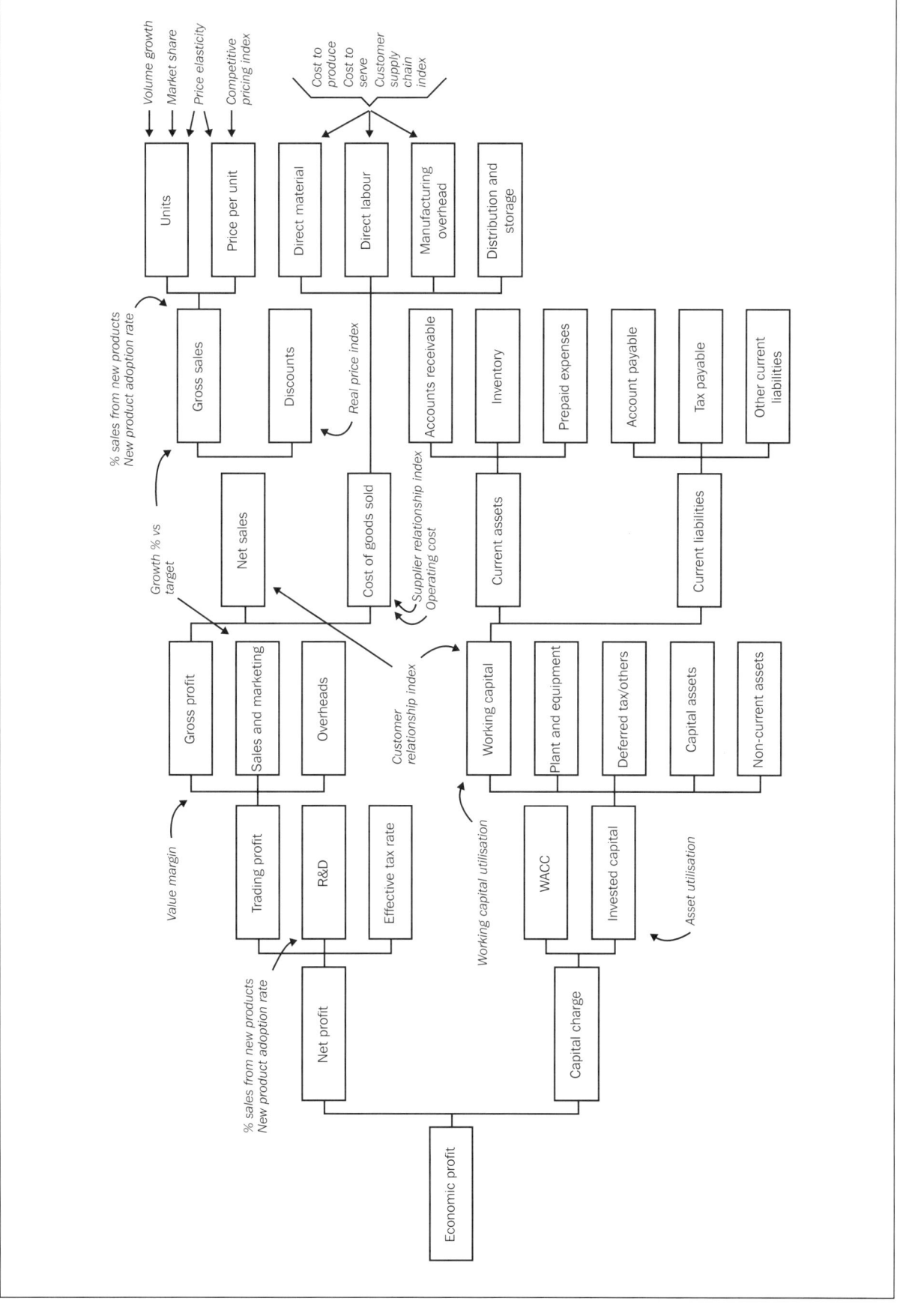

6

Aligning and embedding VBM within the business

CHAPTER SUMMARY

- The last chapter gave guidance on the type of internal performance measurement framework required in order to help give an internally consistent focus on delivering shareholder value.

- This chapter defines what else may need to change and why in order for this type of measurement framework to have real and lasting impact.

- It focuses on the challenge of aligning the measurement framework well within a full business context among a number of business components, such as the performance management review process, rewards and culture.

- In total, there are eight key business components that need to be considered. Their characteristics may need to be changed and fully realigned or redesigned over time. Misalignment of the components means that the business performance management approach is broken.

- Failure to address any one of the components in a bespoke way, in a specific context, will weaken the effectiveness of the overall 'system' or approach. This is because, like any 'system', it needs to be in balance for it to be effective and it is only as good as its weakest link.

- Checklists are therefore given for each of the eight integrated components. These give the key considerations on what to look out for, with a view to focusing on how to (re-)create an effective design.

- Guidance is also given on how to use the checklists through an 'assessment and implications' pro-forma that helps to focus on specific action plans for improvement.

- Going through the full set of components and thinking process will enable the effective alignment of all the components to embed an effective VBM approach, particularly when implemented in keeping with the guidance of the next chapter.

In the life-cycle of every organisation, its ability to succeed in spite of itself eventually runs out.

Anon.

HOW TO ALIGN THE BUSINESS TOWARDS VALUE DELIVERY

The previous chapter defined the type of measurement framework that is required to focus on and cascade a meaningful and integrated value-based measurement approach. It provided guidelines for creating an integrated measurement and information framework to help drive and deliver shareholder value in the context of VBM. Companies that use these guidelines will produce a cascaded set of value-based metrics and a balanced scorecard.

This chapter focuses on designing an improved approach to managing value and performance. It aims to provide a route map that can be used during the implementation phase which is the subject of the next chapter.

So, before looking at the implementation challenge, this chapter reviews all the components of a performance management system that will need to be considered. The key question here therefore is:

What are the essential components of good business management, and how do they fit together in an integrated way to give the maximum desired business impact measured in value terms?

The following sections answer this question by:

- identifying all eight key components that need to be addressed in an integrated way;

- highlighting the key challenge of alignment;

- giving checklists for each component covering the type of questions that need to be answered in building that alignment and delivery of VBM.

The chapter concludes with a series of checklists, one for each of the eight components. These checklists are defined for use by companies which wish to:

- review their current position;

- evaluate the strengths and weaknesses of that position;

- assess the implications of where they are versus where they need to be for delivering successfully against their value-based strategy;

- design the new business characteristics required to adopt, align to and fully embed VBM.

THE EIGHT KEY BUSINESS COMPONENTS

Figure 6.1 shows the logic flow between each of the eight business components that need to be managed and aligned to implement VBM. The logic flow can be best understood by thinking through the following sequence of eight questions:

- *Strategy*:

  What is our approach really trying to achieve and what value will it create?

- *Measures*:

  Does this imply that we need to make changes to our performance measurement system? Are we able to produce the information that we need to support strategy realisation and the delivery of shareholder value?

- *Organisation/structure*:

  Will any accountability changes need to be made as a result of introducing new measures and eliminating some old ones? How and why?

- *Process and stewardship*:

  How is the new measurement and information framework going to be managed to leverage its value and what communications do we need to get people to respond positively?

- *Technology*:

  What implications are there for technology to ensure the right information gets to the right person at the right time?

- *Skills and competencies*:

  Do we need to develop new skills and competencies to leverage the new information that has become available?

- *Rewards and recognition*:

  Do the changes in measures need to be aligned better or differently to individual and team compensation structures? How?

- *Culture and communication*:

  Will the measures result in the behaviours and culture that we want, and how can we plan for all the communication and cultural issues that need to be addressed?

This is powerful stuff. Thinking through these questions, and understanding how to answer them effectively, is a far cry from traditional approaches to improving management information systems (MIS). Too often, a high profile MIS project focuses on defining a new reporting pack that results in something that is merely a bit more 'glossy' with a couple of new measures and graphs. This is hardly a valuable service to the executive directors. Such limited approaches never deliver the impact intended to help improve business performance.

Ready-made solutions do not exist. Bespoke tailoring is necessary. The checklists and design criteria introduced later will help to guide companies that are implementing VBM in order to create that tailored solution. However, some of the complexities and links between the components are worth exploring a little more in order to ensure that the eight components are aligned well within the business.

Fig. 6.1 The eight key business components

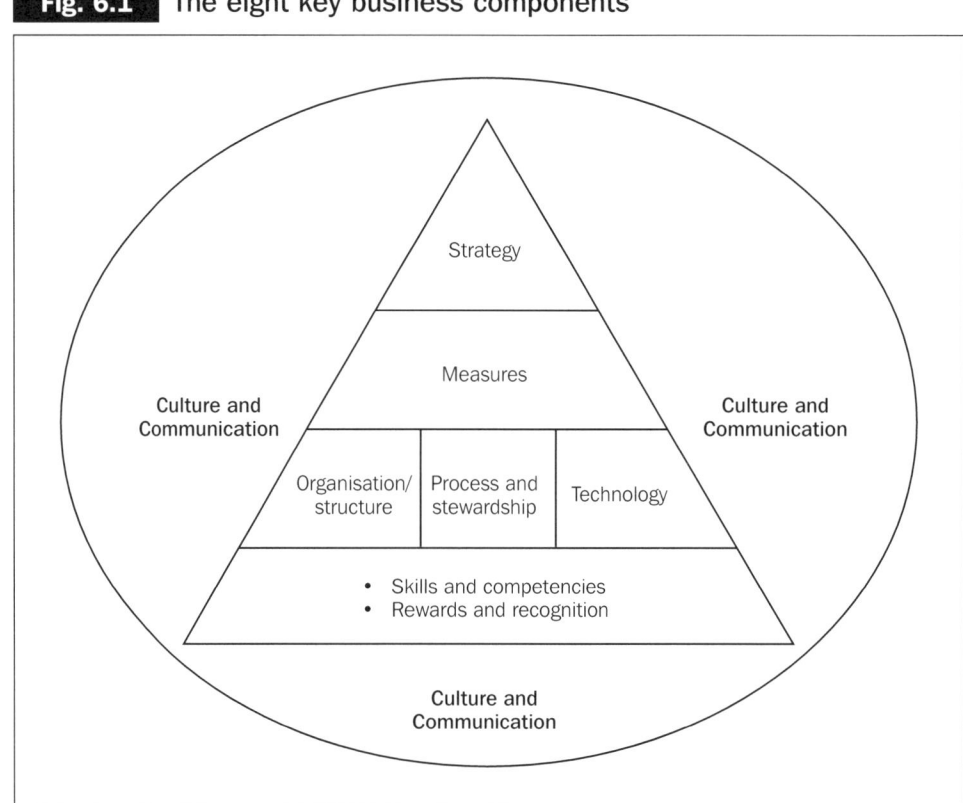

THE VITAL LINKS BETWEEN THE EIGHT BUSINESS COMPONENTS

There are clear interrelationships between the eight business components. The high level of interdependency means that they could equate to a 'business system' in their own right. As a result, in order for the 'system' to work, all the eight system components must work in harmony together. Over time, they may well fall out of alignment as the business focus changes. Therefore, the importance of 'realignment' of all the system components is key and cannot be overemphasised.

Some people see the 'system' as having the potential to create competitive advantage in its own right. The measurement and information framework alone is not enough. It needs to be supported by, for example:

- changing and clarifying key accountabilities;

- rewarding people differently;

- giving them the skills and processes to operate differently;

- putting in place the appropriate processes and technology to enable change to happen.

A chain is only as good as its weakest link. This is also true for the system that these eight components comprise. For example, if a company decides to change its strategy, then it must also change some of the other business components. New measures will need to be introduced and there will also be a need for an aligned reward policy to help drive the appropriate behaviours. Unless this is done, the business system will be misaligned and the new strategy will stand less chance of being achieved successfully.

EXAMPLES OF MISALIGNMENT

Let us consider this point a little further by looking at two more examples. One is focused on the remuneration issue mentioned above and the other at a more general level.

One seasoned sales professional's judgement illustrates the issue on remuneration perfectly. His organisation had operated very much on the basis of regional profit centre accountability for many years. Each region had an administrative reporting relationship to an area office, but the real power lay in corporate HQ. His area was trying to shift focus towards becoming one fully integrated autonomous economic profit centre. The new focus was to be on creating maximum economic profit.

His comment clarifies the strong need for alignment of the new business strategy with the approach to performance management, measures, behaviours and rewards:

I know that continuing to pursue market share in my region alone is not best use of our limited discretionary area marketing funds. My region has reached, and now gone well beyond, the theoretical saturation point for our brands. New markets and regions within our area of operation must offer potentially much more attractive returns and dramatic growth prospects. But what am I to do? Until this company is prepared to change my compensation package that is tied simply to my region's market growth and percentage share, I'm going to fight tooth and nail to keep on increasing my marketing budget allocation!

Clearly, this is a case of misalignment.

The next example is more holistic in nature. It considers a major utility provider that was seeking to shed its traditional image to become fully focussed on maximising economic profit and optimising shareholder value.

The first step in the process they adopted was to carry out a diagnostic on each of the eight business components, recognising that they were out of alignment with their new business strategy. Figure 6.2 illustrates diagrammatically the degree to which they had fallen out of alignment.

Fig. 6.2 Highlighting misalignment and the need for realignment

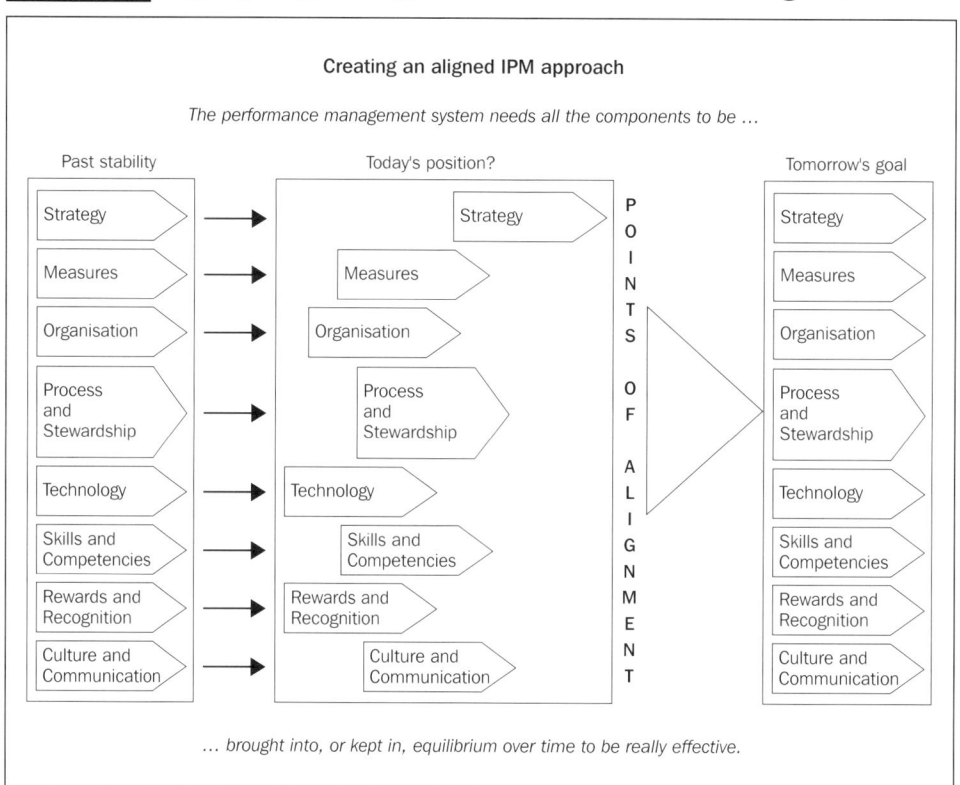

By using the diagnostic analysis and results, they concluded that:

- as a utility provider their eight business components had become misaligned;

- they did not have the enabling systems components focussed on the delivery of their economic profit targets;

- the internal performance measurement system would not support their shareholder value goals;

- they had a broken approach to managing business performance.

Having identified the problems that they needed to address, they set about designing the changes that they needed to make. These changes were analysed separately for each of the eight business components and an extract is shown in Figure 6.3.

Fig. 6.3 Clarifying the challenge of alignment

Component	Change From	Change To
Strategy	Lowest possible cost of operation	Maximum economic value created
Measures	Cost variance to budget	Absolute economic profit contribution
Organisation	'The centre' or 'HQ'	Zonal managers/directors
Process and stewardship	Monthly reports Monthly meeting forums	Closed loop review system Individual management action plans, with consequence management
Technology	MIS and spreadsheet analysis	Full range of decision support tools
Skills and competencies	Financial control and discipline	Value creation and teaming
Reward and recognition	Little discrimination bonuses	Significant positive and negative bonuses
Culture and communication	Passive, reactive	Entrepreneurial

Clearly, this is a very simplified extract of the output produced, as many features were identified within each component that needed to change. They also identified the need for the alignment of each of these components properly. For example, they could not easily have sustained their passion for creating 'Maximum economic value' just by continuing to measure 'Cost variances to budget'.

USING CHECKLISTS TO MEASURE AND BUILD ALIGNMENT

The remainder of this chapter consists of a series of eight checklists, one for each of the business components that have been identified. The aim of these checklists is to help companies implementing VBM to measure and build alignment.

Having used the checklists, diagnosed the situation and defined the design criteria for moving forward, companies will need then to generate specific and concrete action plans to implement the changes required. Figure 6.4 shows a simple blank template for helping to build a bridge between the assessment and the action plans. Figure 6.5 gives an illustration of the type of detail to be captured on the pro-forma that will help to achieve this for one of the components – the process and stewardship component.

Once the changes that are needed to align these eight components have been identified, the next challenge is to implement them. This is the subject of Chapter 7.

Fig. 6.4 Assessment and implications pro-forma

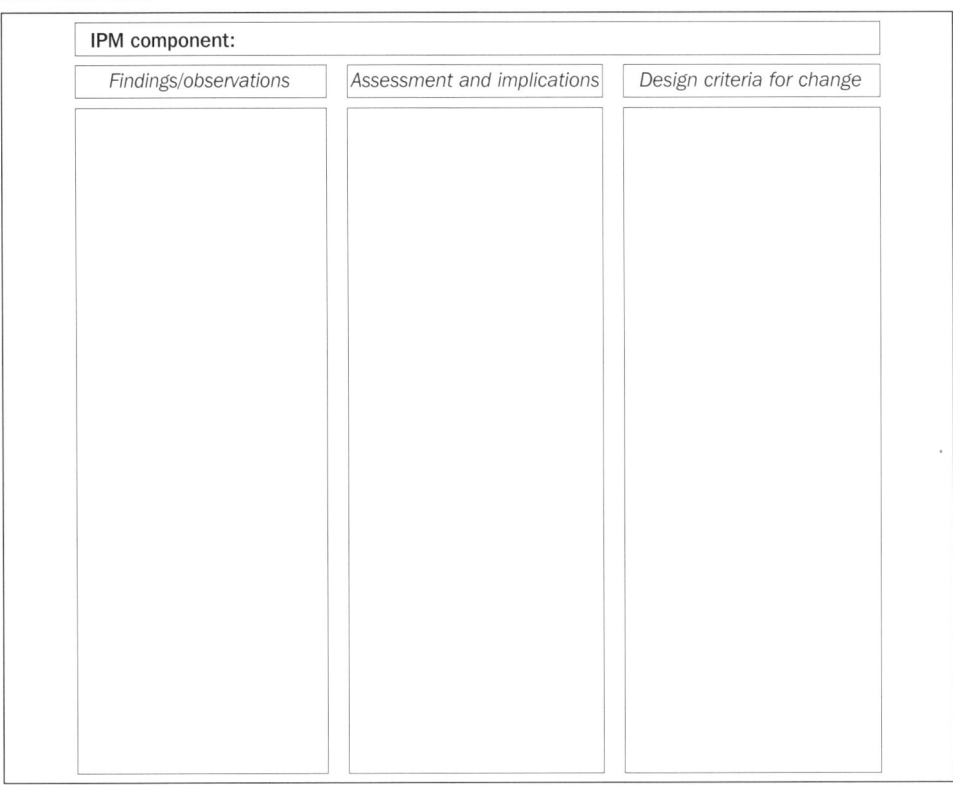

Fig. 6.5 Assessment and implications – illustrative extract

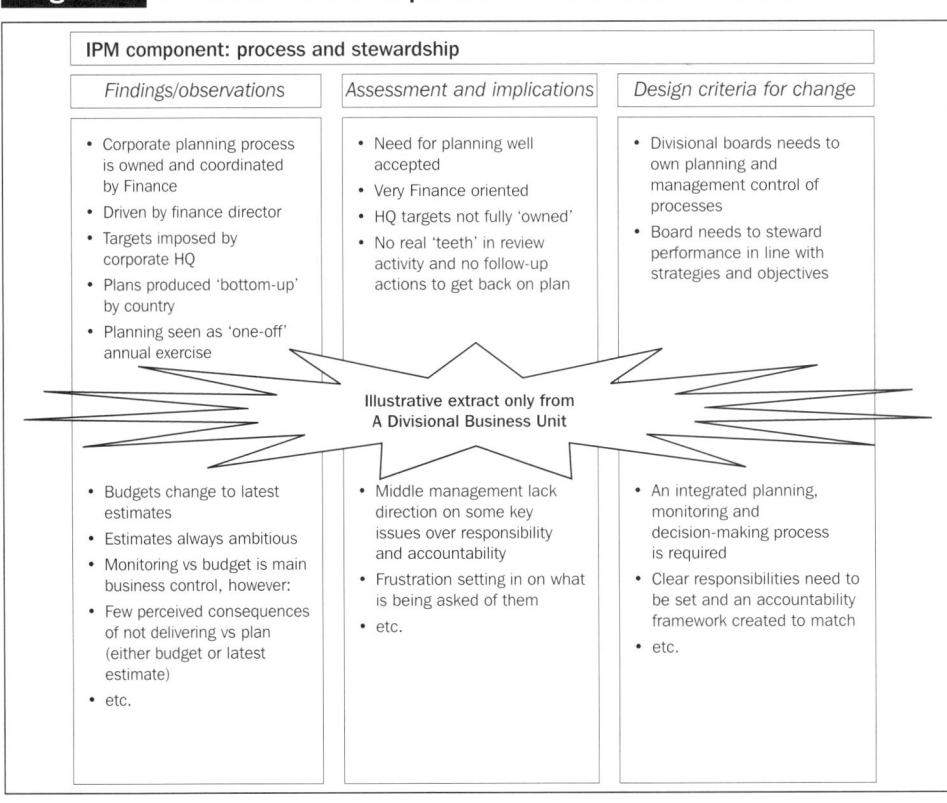

KEY QUESTIONS AND DESIGN CONSIDERATIONS
BY COMPONENT

In using the following checklists in association with the blank pro-forma shown in Figure 6.4, one needs to consider:

■ any significant paradigm shifts that the organisation needs to recognise and respond to;

■ the detailed findings about what is actually happening today;

■ the implications of those findings;

■ the essential design criteria that must be adopted – or rather be met – in order to move the organisation forward consistent with the strategy that drives it.

Checklist 6.1 Strategy

Key questions

■ What are the top 5–15 strategies and related critical success factors (CSFs) at board level against which success can be measured?

■ What decisions are made at board level?

■ What information is requested to support these decisions?

■ What are the individual executives held accountable for – budgets, volumes, profitability, market share, etc.?

■ How is each board agenda established?

■ How is the improvement agenda set – by committee, based on results, or established by each function?

■ How widely are they shared and communicated?

■ Are there any barriers to the full implementation of VBM?

■ Is there a corporate planning process?

■ How does it work?

■ How are the budgets stewarded during the year?

■ Is there a budgeting process?

■ How does it work?

■ How does the corporate planning process fit with the budgeting process?

■ What is the decision process for establishing agreed plans? Is it formal/informal?

■ What decisions are made at the 'mega-process' level?

■ How widely available is information made?

Continued ...

- What prompts individuals to be promoted to senior management positions within the financial organisation?

- What are considered to be the most important communication flows of financial performance?

- What value adding activities does your financial organisation concentrate its efforts on?

- How frequently is business performance subject to a formal, regular review?

Design considerations

- What will your prime measure of business success be?

- What degree of risk are you/your company prepared to take?

- To what extent do you want to use VBM to drive performance improvement and value creation?

- How are you going to enable the drive for competitive advantage by using quality information in developing, executing, maintaining and revising strategies throughout the business?

- How important is it to you to ensure the right information is provided to the right person at the right time?

- What is deemed to be an acceptable level of cost of implementation? Why?

Checklist 6.2 Measures

Key questions

- What information is stewarded regularly at the executive board level?

- What financial and non-financial measures are used in the corporate plan?

- What financial and non-financial measures are used in the budget?

- What management information does the finance organisation produce and for what purpose? Are ad hoc requests completed and for what purpose?

- What are the most critical measures or decision variables considered to be? Why?

- What are the top 5-10 CSFs and measures of performance used by each major process within the organisation?

- Has an ideal set of measures or value-based metrics been redefined recently?

- Was a specific methodology used?

- Is anyone using a balanced scorecard in the business today?

- How many of the ideal measures are in place today?

- When was the last time the measurement set was reviewed?

- How many measures require ABC information? Is ABC in place today?

- What information do line managers within your organisation expect?

Design considerations

- Do measures need to drive real value creation rather than, say, control costs or EPS growth?

- Do some measures need to change quickly?

- Are more holistic, process-based performance drivers needed along with a strong external perspective?

- Can you build on existing concepts that may be being developed as stand-alone solutions (e.g. economic profit, balanced scorecard, activity-based management)?

Checklist 6.3 Organisation

Key questions

- What is the structure of the organisation for the top two to three levels of management?

- How many meeting forums do the executives use?

- What are they and what are they used for?

- How many cross-functional teams are formalised within the organisation?

- What are the formal reporting lines of these teams?

- Would you say everyone has a very clear role and set of responsibilities? Why/why not?

- How credible is the financial organisation within your firm?

- How well does the financial organisation support sophisticated scenario evaluations?

- How would you best describe the financial personnel in your organisation?

- How do you perceive the role of the most senior financial people in your company?

Design considerations

- Is a new and integrated way of working and managing the performance of the business required?

- Is there a need for clear accountability to be established? At what level is there doubt or confusion?

- Who will sponsor/is sponsoring the VBM approach to ensure that the required changes will happen?

Checklist 6.4 Process

Key questions

- What is the current stewardship process at the executive level?
- How are actions created if performance is not as planned?
- How is the executive agenda established and monitored?
- How is performance of the most senior executives stewarded and how often?
- Who is responsible for the current corporate planning process?
- How well does it work?
- Who owns and takes responsibility for completing the budgeting process effectively?
- How is the budgeting process used? How often is it updated?
- How well does the budgeting process work?
- What are the perceived shortcomings of the existing performance monitoring processes?
- How do they manage or reconcile both a process and a functional view of performance?
- How does finance support performance monitoring?
- How is actual performance against financial and non-financial measures monitored?
- What analysis is carried out on a regular basis to understand actual performance?
- What is the current stewardship process below the executive team?
- How are actions created if performance is not as planned?
- How is an improvement agenda established and monitored?
- How do we know the executive management processes are operating as they should?
- What are the key control processes used by the board to manage the organisation?
- What is the most important reporting cycle in your business?

Design considerations

- Who needs to own the planning and management control process?
- Who needs to actively steward performance?

Continued ...

- Who needs to ensure performance levels are in line with expectation or that remedial action is taken?
- What style will be adopted?
- How can a role model of how the stewardship process works be driven from the top?
- Can an integrated planning, monitoring and decision-making process really be adopted?

Checklist 6.5 Technology

Key questions

- What decision support tools are available to the board?
- How are the various scenarios for decision-making generated?
- What technology currently exists to support the monitoring of business performance?
- How well does current technology support the performance management process?
- Do executives have access to an enterprise-wide or executive information system (EIS) system? What is it used for and how?
- What value would an EIS system add to your company?
- How well integrated is the management information within your company?
- How many of the senior managers and executives use PCs on a regular basis?

Design considerations

- What technology solutions can be introduced to support more consistent information collection, reporting, analysis and modelling?
- How can software solutions be made 'user friendly'?
- What are the timescales for delivery?
- Who needs to drive the design work?
- Will the executive reports form the basis for enterprise-wide reporting and communication of performance issues?

Checklist 6.6 Skills and competencies

Key questions

- Have the skills, competencies and knowledge been identified that are required to implement and manage the business plan?

- Has the organisation recognised and identified the 'core' skills and competencies required by the business?

- To what extent has an inventory of skills and competencies been established across the organisation?

- How was this developed and for what purpose?

- How is it updated? With what frequency?

- Is it linked to performance assessment?

- Does it cover business management skills? How?

- Have the skills and competencies required for implementing and maintaining an effective performance management approach been taken into account when designing the competency set?

- What in-house facilities are there for training in advanced performance management or VBM?

- What training has been given to support good performance management or VBM practices over the last 6–12 months?

- When managers change jobs within the organisation, are their development needs identified?

Design considerations

- How can training and development needs be assessed in line with business strategy and the VBM and IPM requirements?

- Who will develop the plan for closing the gap on the skills required relative to those in place?

- How far down the organisation do employees need to be trained and developed to meet the needs of the business?

Checklist 6.7 Reward and recognition

Key questions

- How are the board members rewarded and how is it tied to performance?
- How is team performance recognised and rewarded?
- What is the stated policy on recognising and rewarding performance with respect to the top three levels of management?
- What criteria are used to recognise and reward performance?
- Do these need to change to facilitate the effective realisation of optimum shareholder value?
- How is good performance and exceptional performance recognised and rewarded?
- How is it funded?
- How often is payment made?
- Who determines the amount of performance reward paid to individuals?
- What has been the average payment made over the last three years for each senior manager and executive?
- What percentage of the total compensation package is variable?
- Is the current remuneration system well understood by managers?
- Is the current compensation policy appropriate for meeting the business strategies?
- Do managers automatically receive annual inflation increases?
- How is the total compensation package made up?
- Are there longer-term rewards linked to long-term performance goals?
- Are managers eligible for any non-financial rewards? What are they?
- How is performance and reward compared across the business/countries and between peer groups?
- How are individual and team objectives set for the first three levels of management?
- How is achievement of the objectives assessed, measured and rewarded?
- On what basis does the promotion system work?
- How is this viewed by the managers?
- Does it need to change to deliver maximum shareholder value?
- Do employees receive regular information of the organisation's/business unit's performance against targets/goals?

Continued ...

Design considerations

- What aspects of reward and recognition need to be redesigned to achieve the changes required?

- In particular, is there a readiness to make variable pay a significant element of total pay, based on results?

- Who will be responsible for alignment with performance measures and over what timescale?

- Will short- and long-term performance be recognised by short- and long-term rewards?

Checklist 6.8 Culture

Key questions

- What behaviours are driven by the current measures within the business?
- To what extent would you say that the culture in your company is:
  - hierarchical and functional?
  - closed and ethnocentric?
  - procedure driven?
  - focused on the individual?
  - paternal?
  - risk adverse?
  - entrepreneurial and risk taking?
  - quality driven?
  - customer focused?
  - achievement orientated?
  - open and friendly?
  - energised and empowered?
- What other adjectives would you ascribe to your company?
- Is the culture changing? Why and how?
- How would you describe employees?
- To what extent does the description differ between senior management and more junior levels?
- What are financial staff renowned for?
- What do the top financial staff emphasise most?
- What is your company's attitude to risk?
- How does your company define success and does it celebrate it?
- To what extent does nationality affect company culture?
- Is the culture the same at all levels of the organisation?
- Is your company proactive in trying to change its culture?

Design considerations

- What cultural characteristics would enhance the performance of this company?
- Can they be created? How? Who will take responsibility for bringing about the required changes?
- Is there a passion about creating and delivering shareholder value?

7

Implementing VBM

CHAPTER SUMMARY

- This chapter focuses on the challenge of implementing the types of approaches highlighted so far in this book. The first issue is to recognise or evaluate the extent to which there is a critical business need to adopt the principles within a VBM approach. Clearly, without this need, there is little point in proceeding.

- If the business need is established, VBM initiatives are often project based, at least in the initial stages. What could go wrong? Based on both experience and research studies, this chapter defines the success criteria that need to be borne in mind in order to help build and maintain momentum behind the implementation plan. In particular, this includes the need to fully define the:

  - size of the implementation challenge;

  - essential details of the development approach;

  - detailed implementation plan required.

- It also gives a checklist for project managers on the keys to enduring success, and concludes with a list of top tips on implementation 'dos and don'ts' across the full three VBM perspectives.

It should be borne in mind that there is nothing more difficult to arrange, more doubtful of success, and more dangerous to carry through than initiating changes to a state's constitution. The innovator makes enemies of all those who prospered under the old order and only lukewarm support is forthcoming from those who would prosper under the new.

Niccolo Machiavelli

STARTING VBM AS A PROJECT INITIATIVE

As the first six chapters will have highlighted, addressing all the facets of an integral VBM approach at the same time is no mean feat and can be seen as quite daunting. Indeed, many companies have come to realise that deploying VBM across the full organisation is a multi-year task. This is why taking a VBM approach forward in incremental stages or waves of implementation is often the most common implementation approach adopted.

As a result, the early stages are also often referred to as being the start of a journey. Then, over time, what starts out as a project initiative, often ends up as an integral way of continuously improving the way in which the business is run.

However, unless each stage or wave is addressed and aligned as part of an integrated implementation programme or project early on, there is a substantial risk of failure in realising the full potential benefits over time.

Further, before a project is commissioned or a journey started, an organisation needs to recognise that it has a critical business issue to address. VBM should never be started because it is conceptually attractive or because it might be a nice exercise. This approach is doomed to failure before it's really started.

In considering, therefore, whether a company is ready for a change in approach to the way in which it manages its business performance, it is important to consider how ready it is for change.

Any organisation can only be in one of four modes in its attitude towards changing its performance management approach and adopting VBM. These four modes are as follows:

- As a result of *growth* potential, new challenges need to be managed and it is more likely that a measured change in the approach to managing performance will be required.

- As a result of being in *trouble* or in need of a performance turnaround, some problems need to be more effectively managed and it is more likely that a fast change in the approach to managing performance could be required.

- As a result of some *stability* with little market turmoil, existing approaches to managing performance could be considered to be quite satisfactory.

- As a result of being *overconfident*, it is unlikely that any change in the approach to managing performance would be contemplated until it is too late (or because the approach is genuinely superb!).

In other words, it is only when an organisation recognises that it is in one of the first two states above that it will seriously consider making a substantive change in approach towards the principles articulated under the VBM design criteria. In reality, it is difficult to see why most organisations would not find themselves in one of these first two categories, however, due to the explosion of the e-factor and increasing globalisation!

READY FOR A CHANGE IN APPROACH?

When the need to change has been recognised, an effective change management process will help to demonstrate how critical and urgent the proposed change is in response to the key drivers or pressures for change. Figure 7.1 identifies a series of questions that calibrate the degree to which an organisation is ready for the implementation challenge. They are primarily concerned with clarifying motives, understanding whether there is a sufficiently compelling case for change to see it through the difficult period of transition, and testing out the clarity of what it is trying to change towards.

Fig. 7.1 Change readiness assessment questions

Cost of no change	• Is the need to implement the change accepted as a business imperative? • What sense of urgency for the change is felt throughout the organisation? • How aware of the existing 'pain' is the organisation? • Has the business case for the change been communicated?
Ease of change	• Have the scale and scope of the likely changes been assessed? • Has the impact on the organisation been evaluated? • Have the required quality of resources been made available to ensure change is delivered? • Do leaders have the knowledge, authority and leadership skills to take people along with them?
Clarity of way forward	• Is there consensus on the organisation's strategic vision for a new approach? • Is there awareness of the potential gain from achieving the vision? • Have business objectives been agreed with all stakeholders? • Is it clear how the project will help deliver the organisation's strategic vision? • Is it understood how this initiative complements others?

Before agreeing on any VBM project, therefore, the organisation should assess the impact of not implementing the proposed changes. In particular, it should identify the effect or cost of any delay to their implementation. This is a good general control procedure for approving and prioritising any type of project. It is also vital for communicating the need for change within the organisation and to get essential buy-in to VBM. The questions within Figure 7.1 will help to support these requirements and are the logical starting point for justifying any new move towards adopting a VBM approach.

THE IMPLEMENTATION CHALLENGE

On reaching a conclusion that a VBM approach will help address key business issues, it is important to move on quickly to the detailed project implementation planning stage. Based on the contents of Chapters 1 to 6, combined with perhaps other research and materials, some considerable effort or 'pre-thinking' will have already gone into defining the need for an improved business performance management approach and what changes may need to be made. The outputs from this process of preparation should therefore now have delivered to hand:

- the burning business issues that are forcing the need to consider a new approach;

- a good understanding of what is now the 'art of the possible' and what leading companies are now striving towards;

- the needs and benefits that are being sought from the new approach, at least at a high level;

- a detailed understanding and assessment of the current position;

- the ability to benchmark against these alternative positions;

- a strong view of the obstacles and barriers to making progress.

This chapter now goes on to highlight how to leverage those outputs as inputs into implementation planning in order to make rapid progress. Often, a 'kick-start' workshop, or series of workshops, can be a useful process to help engage people in really starting to think hard about changing the way business performance is being managed and why it is important to them. The success measure for this process is that they then want to make a difference and believe that they can. It achieves it by creating energy and commitment to focused change through clarifying the benefits of changing, and the urgency of the need to do so, along with an in-depth understanding of what will be involved.

FIFTEEN PROJECT MANAGEMENT KEYS TO VBM SUCCESS

Good planning is a prerequisite for successful project implementation. However, many projects commence with little more than a start and finish date and a vague idea of what will happen in between. The output from the 'kick-start' workshop(s) can be a rich input into the creation of a soundly based and comprehensive project charter or plan. This plan will need to ensure that it:

- clearly identifies the goals, objectives and anticipated benefits of the initiative;

- specifies the precise deliverables from it;

- assesses the level of risk of failure and readiness of the organisation to accept the changes (see Figure 7.1);
  (*Note*: Cultural acceptance is key and needs to continue to be assessed as you go through implementation and a learning curve. Checkpoints can be set to test the degree of progress and acceptance being made – see Figure 7.3 as described later on in this chapter.)

- crystallises any significant obstacles or hurdles to be overcome;

- explains and clarifies any assumptions made;

- defines a detailed work plan at an activity level – this clearly needs to be at a much greater level of detail for the initial stages than for the later ones;

- estimates the resources required and who needs to be involved at which stage and why;

- defines the criteria for judging the success of the project at each stage.

As you will have noted by now, a 'kick-start' workshop may clearly help in creating much of this planning content in a way that is already building up awareness, buy-in and commitment among those who matter most in progressing the endeavour.

The earlier sections covered why the definition of the urgency for change and the business imperative were so important. Based on extensive implementation experience and some research studies, the other success criteria to maintain momentum can now be defined very clearly. For ease of reference, these are shown in Figure 7.2.

Fig. 7.2 Fifteen project management keys to VBM success

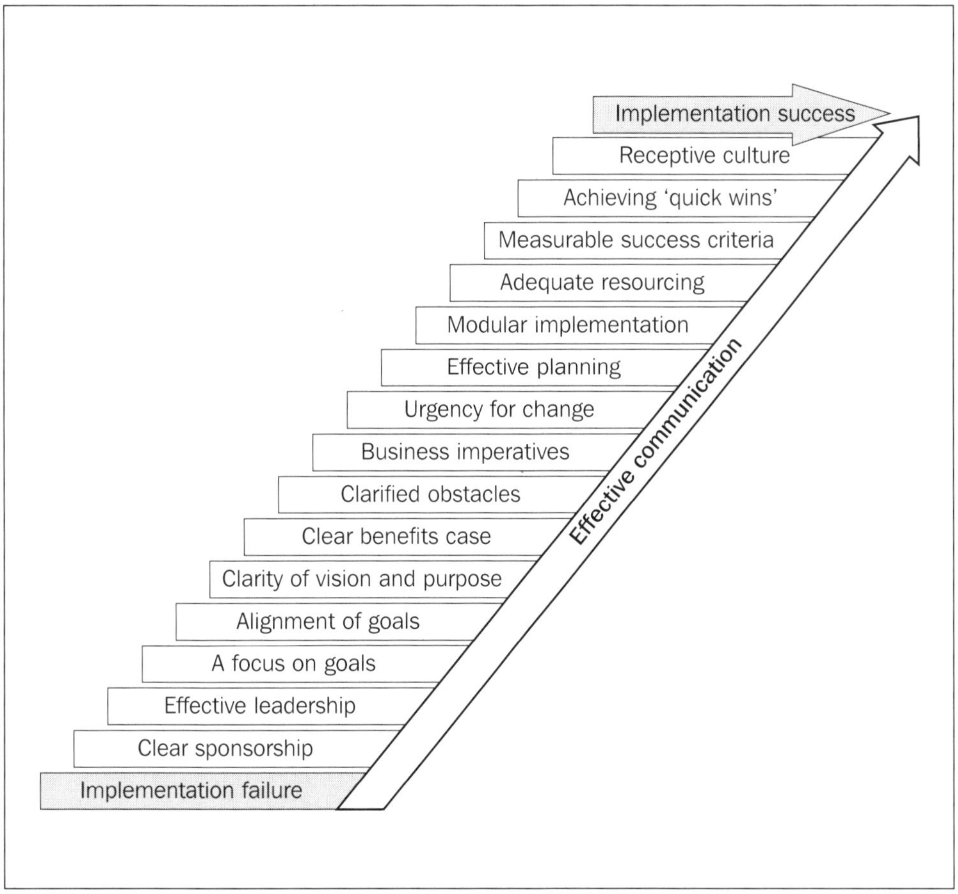

Many of these success criteria will, by now, be self-evident. However, a number of the items within this figure are worth exploring a little further.

Clear sponsorship

This almost needs no introduction. Lack of clear sponsorship is the most frequently quoted reason for any type of project or initiative failure and this is particularly true of VBM. Don't forget to make sure that clear sponsorship is present in abundance and can be retained throughout.

Aligning the initiative with organisational goals

Any project must have a clear role to play in helping to achieve organisational goals. The key question to be answered before pursuing the implementation of a new VBM approach is:

How will it help in achieving the vision, strategy and objectives of the organisation, and how will it play a part in both achieving success in its chosen marketplace and delivering superior shareholder value?

An excellent VBM approach will actually be designed to help drive future performance and the realisation of strategic intent by helping directly make decisions, manage processes, and measure progress through:

- understanding the true sources of profitability in existing and potential markets;
- managing resources effectively through critical processes;
- linking goals to activities so that employees at all levels know what is expected of them and how their contribution will be measured;
- providing information easily to the right person, at the right time and in the right place.

Effective leadership

While this might seem an obvious requirement for any project, it is important to understand the distinction between sponsorship and leadership. A sponsor might approve and lend some support to an initiative, but a leader takes responsibility for the change, and is personally committed to it.

For successful transition to a new environment, visible, high-level leadership will be required, supported by continuous demonstration of how the changes implemented have delivered/will deliver value to the organisation.

Without this visible leadership at the outset (or at the appropriate stage of prototype development) the VBM project is doomed to failure.

Modular implementation

VBM is like an elephant. It can only be eaten in 'bite-sized' chunks. In other words, VBM is a multi-year, multi-faceted challenge. Do not try and do everything at once. Break the implementation plan into three six-month modules within an overarching VBM vision.

Adequate resourcing

The detailed implementation plans developed for the project should help to identify the resource requirements at various stages. The required skills for each of the tasks should also be identified.

A common mistake is to plan the requirements for a central project team in great detail but to pay lip service to the input required from operational areas of the company affected by the project. In addition, in systems projects many weeks are often spent specifying the design of a new application but then the users are expected to review and sign off their acceptance of the system in a matter of a few days. Are these resource implications familiar?

A good rule to apply is not to proceed if adequate resources for the proposed stage of development are really not available. If you proceed and fail to deliver quality results the reputation of the project will suffer. However, don't let this be an opportunity for people intentionally to block planned developments. It is never the right time because there are always too many other things going on. Therefore, take on board the views of management and staff and modify plans accordingly. Get commitment to a start date if the keys to success are achievable, but perhaps over a slightly different timescale.

Receptiveness of the culture of the organisation to change

Regardless of the type of project, an approach that works well in one company may hit severe difficulties in another. The shared values that drive the collective behaviour of individuals will vary from one organisation to the next. Those that have very demanding and high achieving cultures might welcome initiatives that help them improve still further, despite the effort required. Other companies may be very defensive and wary of any change, and this may take a passive or aggressive nature depending on what is normal and acceptable behaviour in the workforce.

The type of culture existing within the organisation should therefore be taken into account when developing VBM implementation plans. These should then include specific change management action plans designed to overcome any expected resistance to change, or difficulty that individuals may have adapting to new ways of working.

The organisation's training and development programme can play a major part in shaping people's attitudes and the way in which they behave at work. Training can also be used to provide people with information on the corporate vision and strategy, the reasoning behind it and how VBM will help. This sharing of information is a vital part of empowering any workforce – if you want people to change their behaviour and accept greater responsibility for corporate performance, you must provide them with information to understand the objectives that have been set and the importance of achieving them.

In addition to consideration of the organisation's culture, measures can be applied to judge whether an organisation is prepared for change. A number of questions can be asked as part of a basic 'change readiness assessment', and examples have been shown in Figure 7.1.

If the answer to any of these questions is vague or negative, appropriate action should be taken to rectify the situation.

Full and effective communications

A well-constructed communication plan is integral to the implementation process for any major project and this can fundamentally underpin success, as suggested in Figure 7.2. Good communication is also essential throughout and the objective must be to support the movement of people up the scale illustrated in Figure 7.3. Success must be measured not by people's awareness of the initiative but by their commitment to the changes that will occur.

Fig. 7.3 Building the commitment curve

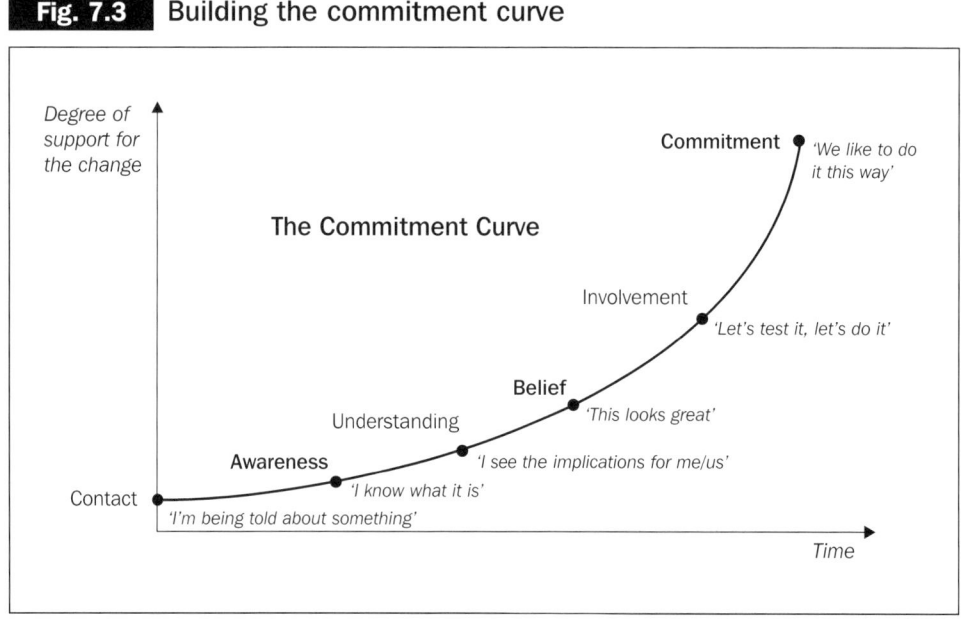

Remember there is more to communication than simply issuing internal memos about proposed actions. Good communication is 'ongoing' not a 'one-off' and is aimed at gaining true understanding and a two-way dialogue with people in the organisation. In order to succeed, a range of communication techniques across a range of media is needed, for example, presentations, workshops, briefings, videos, publications.

The success of the communication strategy can be measured by applying the ABC diagnostic suggested in Figure 7.3 that is focused on understanding whether there is:

A – awareness of the need and potential benefit;

B – belief in the deliverables;

C – commitment to making it happen.

You can get insights as to where the company is on this scale by judging people's reactions to the progress being made.

Really effective communication in this type of exercise is clearly dependent on *visible and early signs of success*. Examples are often found in the initial piloting and prototyping stages and this is why taking this kind of approach can be a vital source of maintaining momentum, even if it is not generally considered fundamental to the success of the overall initiative.

IMPLEMENTATION CONCLUSIONS – DOS AND DON'TS

To conclude this chapter, a number of practical lessons or observation from the 'front line' are summarised in a series of concrete 'dos' and 'don'ts' that add further substance to securing a successful development and implementation approach. They also act as a helpful checklist of top tips.

These dos and don'ts are based on extensive practical experience of implementation support, research, training and management education. They are categorised under the following two key headings:

- the first deals with general project and change management issues on implementing VBM – project and change management: dos and don'ts;

- the second deals with the content and structure of the VBM approach and how it can best be addressed through the project plan – VBM structure and content: dos and don'ts.

Project and change management

Do

- Ensure that there is a business imperative for a new performance management approach. Make sure that this is well recognised by the organisation before implementation proceeds (too far).

- Consider the use of a pilot study to demonstrate how excellent management information will help the organisation.

- Ensure that the vision for the new VBM approach is aligned with the organisation's strategy.

- Demonstrate how the new approach will align improvement initiatives within the organisation by focusing them on common value-creating objectives.

- Recognise that the transition from financial analysis and reporting on the business to implementing a demanding VBM discipline and culture requires a change in the leadership style.

- Remember that high-level leadership of the VBM project is required.

- Consider a sponsor perhaps from outside the finance function.

- Remember that when developing the means by which individuals measure the performance of themselves and the teams to which they belong, and indeed the means by which they will be measured by others, their input and support is essential to get commitment.

- Consider the culture of the organisation and build specific actions into the development programme accordingly.

- Remember that achieving change can be a severe shock for people who have only ever been asked to work in one particular way.

- Consider changes that may be appropriate to the company's pay and reward systems to align personal objectives with corporate performance measures.

- Consider the training requirements essential for people to gain a full understanding of the development of performance measures and to help them interpret and use the information.

- Recognise that communication is a two-way and ongoing process.

- Be proactive in looking for problem areas and allow time to win people round to the new approach proposed.

- Ignore the significance of communication as a driver of change.

- Take people along with you!

- Plan in detail for the next phases of the programme.

Project and change management

Don't

- Plan in detail for all the phases of the whole VBM programme.

- Proceed with any VBM development until the organisation understands how the implementation will support the achievement of the organisation's goals.

- Allow momentum to be lost by the departure of the sponsor through career progression.

- Forget to plan and agree the resource requirements of operational areas as well as the central project team (and the inputs required from all other parties!).

- Allow resistance within the organisation to derail the development programme.

- Be disheartened by progress being slower than you would ideally like. This is often a fact of life!

VBM structure and content

Do

- Ensure that shareholder value is genuinely embraced within the business and senior management culture.

- Measure it and reward managers based upon it.

- Reflect shareholder value creation within your top level business plans and targets.

- Keep the internal methods as simple as possible so that they are understandable by the average manager.

- Limit the number of managers with responsibility for external investor relations – they will have to cope with high levels of technical complexity.

- Provide them with appropriate support and data tools to develop their expertise.

- Stay on top of the investor community.

- Know what drives your share price.

- Identify and court key potential investors.

- Manage investor communications carefully.

Continued ...

- Be prepared to demonstrate that your chosen approaches work in delivering improved TSR relative to peers.

- Be prepared to be radical in the pursuit of value. If appropriate take one-off actions to increase shareholder value.

- Recognise that share price can be increased, not only by improving ongoing future performance in a managed fashion, but also by one-off restructurings to improve or realise hidden commercial potential.

- Obtain top management support before you implement the VBM approach.

- Develop your own consistent common language for internal communication.

- Build on, rather than throw out, the 'old' measures that are already in place and are understood by managers.

- Develop effective and visual links between the old and the new measures.

- Develop a clear framework for all new measures, with a stated purpose for each one.

- Make sure that this framework shows a clear link between the corporate goal and the value drivers controlled by operational managers.

- Demonstrate this link dynamically and visually with real in-house examples.

- Educate all those involved, both in financial/strategic fundamentals and in VBM frameworks/terminology.

- Limit the use of complex models to those who need to use them to add value to the decision-making process.

- Evaluate strategic options by comparing the future value of each option.

- Cascade VBM down to the level of products/segments/customers, as long as the data is reliable and meaningful.

- Remember that small changes to value drivers can have a large impact on business values.

- Try to find out about the models which external analysts are using and the assumptions they feed in.

- Compare your own valuations of your business to share price and look for the reasons for differences constantly.

- Remember that, internally and externally, much of value creation is about long-term cash flow.

VBM structure and content

Don't

- Forget that many external analysts still use traditional measures like EPS and the P/E ratio.

- Assume too much financial understanding among non-financial managers.

- Get hooked on labels and jargon.

- Assume that terms are always defined in the same way across different companies, analysts, consultants and business advisers.

- Assume that everyone has to know the detail of everything.

- Allow the 'techies' to become obsessed with perfecting the mathematics of the model.

- Cascade complexity down to levels where it distracts rather than directs and supports.

- Allow VBM to be seen purely as a financial process.

- Allow any new measure to be introduced unless it has a clearly stated role in the cascade.

- Believe that there is much exact science in VBM, particularly around the calculation of cost of capital.

- Use market-based shareholder measures like TSR over too short a period or fail to measure against relevant peers.

- Use ROCE or other percentage measures unless they are combined with growth targets.

- Cascade VBM to products and segments unless there is a good ABC-based costing system to support it.

- Worry about publishing links and visual frameworks which seem 'blindingly obvious' to financial people – other managers will appreciate them!

Where is VBM heading?

CHAPTER SUMMARY

- In future, the evidence for the successful delivery of superior shareholder value through VBM will increase. The companies that deliver superior shareholder value will, therefore, provide a blueprint or role model that others will seek to emulate, particularly in the small and medium-sized business (SME) sector.

- In the medium term TSR will probably continue as the main goal and measure of corporate shareholder value.

- Top management will need to continue to tailor their company's performance measurement systems to meet the unique needs of their people and culture. They will avoid excessive complexity in order to achieve a good level of buy-in and understanding throughout the business.

- For internal performance measurement, companies will become increasingly attracted to economic profit. Its complementary fit with TSR and its focus on bringing balance sheet and cash flow variables into the profit and loss account will help to change the mindset of managers. As a result, economic profit will become established as the superior overarching internal performance measure.

- For reporting performance externally, market-driven valuations and metrics will become increasingly important. Companies will need to cope with much greater levels of complexity as stock market analysts further develop their forecasting and valuation models.

- Companies with few assets and high market values will focus on measuring and managing the drivers of 'intangibles', such as goodwill, knowledge and research and development. The importance of 'intangibles' in shareholder value is already evident and this importance will grow at an accelerating rate.

- Companies will need to consider the needs of all stakeholders, not just their equity shareholders, when making key strategic decisions.

- The balanced scorecard will emerge even more strongly as an attractive and complementary framework to economic profit, enabling the drivers of value to be measured and cascaded meaningfully within the business.

- In future, the most successful VBM companies will be the ones that best address a challenging paradox. They will need to:
  - understand a good deal of detailed technical and financial complexity in shareholder value and business performance measurement; and
  - at the same time deliver simple, powerful and effective internal and external communications.

Vision without action is merely a dream. Action without vision just passes the time. Vision with action can change the world.

Joel Arthur Barker

VBM COMPANIES WILL PROVIDE A BLUEPRINT FOR OTHERS

This final chapter focuses on the future. It suggests how VBM is likely to develop, how VBM principles will be used by businesses, and the key challenges that companies will need to overcome to achieve success.

It also considers the extent to which VBM will become fully and permanently embedded, both within the companies that espouse it and also within general business management techniques.

Chapter 1 highlighted several companies that have adopted VBM and built up a good track record of delivering superior shareholder value. This has prompted others to follow suit. In doing so, they have recognised that this management philosophy is not simply a 'new fad' but an approach that has delivered sustainable competitive advantage over a number of years.

The continued successful adoption of VBM in leading companies will help to provide a blueprint or role model that others will seek to emulate. This process has already started. More and more leading companies are benefiting from a focus on VBM. Its adoption may even increase with the arrival of a second wave of VBM practitioners, particularly in the small and medium-sized business (SME) sector, as they seek to benefit from the experience gained in the 'leading pack' of companies.

TSR WILL HAVE A LONG LIFE

TSR is relatively simple to understand and to measure. It has been highlighted as the main corporate goal for most companies who have embraced VBM principles, and even some that have not. As a result, it is likely to remain prominent as the common focus for reconciling shareholder needs to company performance, shareholder cash flow to company cash flow.

Despite TSR's dominance there will remain questions about of its validity over both the short and the long term.

The short-term questions arise because of stock market trends and volatility. Some companies have seen their TSR turn negative as a result of relatively short-term fluctuations in the stock market. 'New economy' shares have swung in and out of fashion as the target of investors, driving down the share prices of many more established companies. The argument is that performance should be seen as relative and long term. However, managers are bound to become sceptical about

TSR's validity if short-term market fluctuations, over which they have little control, are allowed to affect their bonuses.

The long-term reservations about TSR are even more powerful. Some reservations have been expressed about the TSR 'treadmill' and how it becomes increasingly difficult for even the most talented top management teams to keep delivering superior TSR over several periods. There may therefore be a tendency for managers to leave after achieving their first phase of performance improvement. This is because TSR for the next period requires the beating of expectations all over again, from the base of the improved share price. Thus companies may feel the need to move to another metric as a second phase, particularly if the first phase has involved a major turnaround in performance.

Despite these criticisms, TSR is likely to continue as the primary corporate shareholder value goal and top-level performance measure, at least until a better metric emerges and becomes fully established across the business community.

SIMPLICITY AND PRACTICALITY WILL REMAIN VITAL

VBM, like any other management philosophy, needs to be applied in a practical way. Top managers and their financial colleagues need to find out what works best for them. They realise that they should apply, say, TSR or economic profit in a common-sense way when they are evaluating strategies and implementing performance management frameworks. Companies are therefore taking a more practical view about the internal implementation of VBM principles as they move further along the learning curve.

There is a story told in one FTSE company which illustrates the need for practicality. The company had traditionally measured performance by using ROCE. Now, following management changes, they were aiming to introduce economic profit. Therefore a decision had to be made about the cost of capital to be used for the economic profit calculation. External consultants had calculated an excessively precise WACC, of around 10.6754%, and had recommended different levels of WACC for each of the divisions to cover the different levels of risk. There was a meeting to discuss this issue and it went on for some time.

The CEO, calculating the time and cost being wasted and knowing from his MBA studies that the calculation of WACC was something nobody could ever agree upon, stopped the meeting and said:

Ladies and gentlemen, this meeting will stop now. The answer is 10%. Because, if it is 10% I can calculate it easily and so can my managers. And frankly I don't care whether it is technically correct. All I care about is that all my managers see the impact of their decisions about fixed assets and working capital and change their behaviour accordingly. Now can we move on to decisions which will really create value.

The CEO was right to focus on practicalities. His views seem to align with those companies who have embraced VBM principles as a way of managing their businesses, companies like Unilever, Boots and Cadbury Schweppes. The main success stories have *not* generally arisen from the building of excessively complex models. In most cases DCF-based approaches were already used by these businesses anyway. What was needed was a simple value-based approach that managers could understand, but which would nevertheless help to change their behaviours and thinking.

The real changes in mindset have been brought about by the introduction of balance sheet and cash flow management principles into the day-to-day thinking of managers, into their cost statements and into their profitability analyses. It is that extra line at the bottom of the profit and loss account, reminding managers that assets cost money and must be accounted for, which has made the real difference to performance. It is the relative simplicity and focus of economic profit, combined with the use of an internally consistent performance measurement cascade, that has helped change the mindset about true value creation.

The level of complexity and sophistication should be matched exactly to the need. Top managers will need to continue to tailor their performance measurement systems to address the unique needs of their people and culture. They will also need to reflect the needs of different levels of management and decision-making, whether globally or regionally, and whether across business units or within individual processes and functions.

THE ECONOMIC PROFIT TOOL WILL REMAIN POPULAR

Economic profit will increasingly be adopted as the focal point for integrating the performance measurement cascade. Again, however, simplicity will be key. Rather than having to make numerous adjustments to be theoretically 'pure', the emphasis will focus on a simple calculation, one cost of capital figure and relatively few adjustments. It will be designed so that everyone within the organisation is able to understand the logic, completeness and simplicity of the approach.

Economic profit will become increasingly embedded within the internal performance measurement system. Companies will become increasingly attracted by its completeness and its ability to help change the management's mindset through bringing balance sheet and cash flow variables into the profit and loss account. As a result, economic profit will become established as the prime overarching internal performance measure for business units. However, the adjustments made within the calculation may vary to reflect some of the sophistications of the investor community, as discussed below.

THE INVESTOR COMMUNITY MAY ADD COMPLEXITY

The external perspective will be very different from the internal one. In relations with the financial community, we see the present trend towards greater complexity continuing even further. There will be an increasing need to understand how stock market analysts appraise and value companies. With more advanced forecasting models and the new valuation methods being developed, analysts will have different perspectives and new insights for checking their share price targets.

In addition, there will be an increasing recognition that companies have significant value in intangible assets. These may broaden the traditional understanding of items such as goodwill. For example, some argue that good suppliers add value to a company, or that an expanding and high-quality customer portfolio is as much an asset as 'bricks and mortar'. The key issue will be how to value and report such items and how – or whether – to include them in economic profit calculations. If so, they will often have a significant effect on economic profit.

There are also some interesting trends in the techniques and in the data that analysts are using in the share valuations embodied within their published broker research. These include:

- the move to increasing complexity and new metrics;
- the retention of all the older metrics at the same time, which generates an increase in the sheer volume of different measures;
- the increasing use of value-based measures such as cash (FCF, operating cash flow, etc.), economic profit, spread, EV ratios;
- the fact that most valuations are relative to sector and not absolute.

In the early days simple metrics, such as EPS and P/E, enabled us to perform relative valuations within a particular sector or peer group. More recently measures have emerged that reflect growth, discounted cash flows (such as residual income or economic profit techniques) and market-based values (using

EV and EV-based ratios). The previous measures have also been retained so most analyst reports now contain a wide variety of measures.

Strictly speaking, the discounted cash flow and residual income based measures are the ones that most closely identify shareholder value added. However, companies will need to appreciate the full breadth of measures used in order to manage their investor relations effectively.

Figure 8.1 shows how the complexity of valuation metrics has grown. In the future we are likely to see still more measures in use. Real options and Enterprise Value Added (EV+), which we discussed in Chapter 2, are two of the new approaches that may become more widely used.

Fig. 8.1 The growth in the measurement of shareholder value

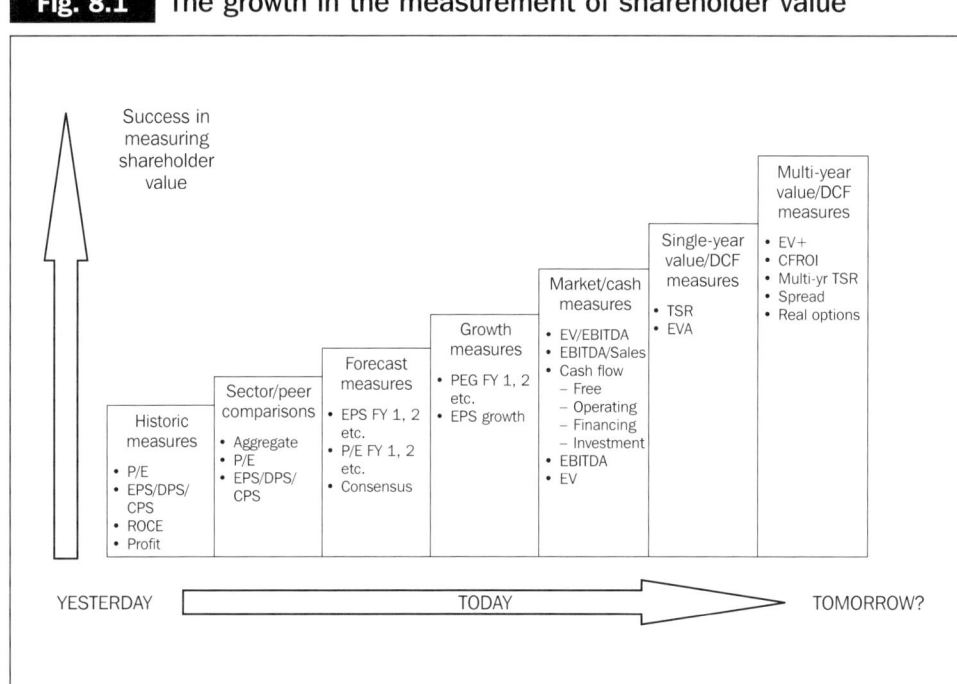

This increasing volume and complexity of valuation metrics is bound to have an impact. It will spread from analysts to fund managers and then to the IR departments of sophisticated companies. Ultimately even smaller companies will see the need to have expertise in these topics, although they may need outside help to obtain it.

Companies will have to become more knowledgeable and impressive when communicating with stock market analysts and fund managers, reflecting an understanding of the increasingly complex models they will have to use. Companies will need their own internal versions of such models and they will need to be ready to mirror the analysts' calculations.

For reporting performance externally, therefore, companies will need to cope with much greater levels of sophistication in the investor community.

THE BALANCED SCORECARD WILL HAVE
A PIVOTAL ROLE

There is a barrier to the wider adoption of VBM among some companies operating in certain geographical areas and in certain cultures. The barrier arises in relation to its association with the term *shareholder*. For example, there may be some rejection of the concept of maximising value for shareholders, particularly in non-western countries. This may be replaced by a general desire to measure value to other *stakeholders*.

It may be easy to relate to the shareholder value concept if you are brought up in a western capitalist environment, but it is not an easy message to deliver in some national cultures. It is also sometimes quite difficult to convince top managers in European and North American companies that shareholder value is not necessarily a motivating concept to managers in, say, Korea or Columbia.

Such rejection is not necessarily confined to managers in these countries. In any business environment, an emphasis on financial performance alone is likely to be challenged, perhaps with some justification. It is true that it can be convincingly argued that the maximisation of shareholder value will also, in the long term, deliver best value to customers, employees and all other stakeholders. But it remains a potentially narrow focus, particularly in the short term. We foresee a continuous and increasing challenge to the exclusive focus on shareholder return.

This means that it becomes vital to tailor the performance measurement framework to the needs of the business. The framework must reflect the behaviours which are desired, the level of complexity which managers are able to cope with and the unique culture which each company and country has developed.

In addition, companies with few assets and high market values will need to be able to understand how to measure and manage the drivers of value in their intangibles. These include goodwill, knowledge and investment in research and development. The importance of intangibles in shareholder value measurement has already started and this will continue to increase at an accelerating rate.

The conclusion of all this is that the balanced business scorecard, discussed in Chapter 5, will emerge even more strongly as a performance measurement framework in the years to come. The successful companies will be those which can retain the essence of VBM thinking, while at the same time developing a tailored, and complementary, balanced scorecard that matches their unique culture and internal communication needs.

MANAGERS WILL FACE A CHALLENGING PARADOX

The future is one in which economic profit and the balanced scorecard will both have a pivotal role. Companies will aim to implement these techniques in a simple and complementary manner that their employees understand. Yet we also envisage that the sophistication of the investor community, and the broader responsibilities that companies owe to stakeholders, will add complexities that need to be managed carefully. So how should top management prepare for a future which might be seen as the VBM paradox: a move to internal simplicity but external complexity?

In our view the paradox only really exists at a superficial level. Clearly, not everyone within the organisation will need to appreciate all of the issues and techniques we have described. However, it is likely that financial directors and key members of their teams will need to master them and communicate effectively with increasingly sophisticated analysts and fund managers. For example, companies will need to be able to discuss the impact of their business decisions on economic profit and share price. If necessary they may have to enter into debates about technical issues, such as the valuation of their intangible assets.

We believe that some of the companies highlighted in this book have shown the way forward. Their simplified and practical internal approaches are complemented by centralised expertise with a complete and rigorous knowledge of the concepts and metrics that relate to shareholder value. These experts have the confidence and the ability to simplify internal measures effectively. They have an in-depth understanding and are aware of the potential pitfalls in any particular approach or technique. They appreciate the numerous interdependent financial and non-financial relationships that need to be understood and managed within an integrated business model across the three perspectives of the VBM framework that we have outlined.

In future, the most successful VBM companies will therefore be the ones that best address the VBM paradox by:

■ understanding the increasing technical complexities in shareholder value and business performance measurement and communicating externally; while

■ operating simple, powerful and internally consistent internal performance measurement tools for communicating internally.

The management teams in successful VBM companies will therefore be performing a clever juggling act. We hope that this book has gone a little way in helping to bridge the gap between these two distinct needs.

Appendices

Discounted cash flow (DCF) techniques

DCF HELPS US TO COMPARE TRANSACTIONS OCCURRING AT DIFFERENT TIMES

If we compare two different transactions that occur on the same day, it is easy to assess whether value has been created. In the unlikely event that we bought a share for £100 and sold it for £150 on the same day, £50 value would clearly have been created. If, however, that $150 is realised in one year or five years' time, there has to be recognition that less value has been created. Money received in a year's time is worth less than money today.

The conversion of cash flows to 'present values' enables us to quantify the worth of future cash flows. Once all cash flows are expressed in present value terms then they can be compared more accurately. First, however, we need to make an assumption about our cost of capital, the percentage by which we need to discount future cash flows each year. For simplicity we will assume that the cost of capital for the shareholder is 10% per year (the factors in the calculation of this percentage are contained in Appendix 2).

The use of DCF techniques enable us to ask the question 'if £100 today is worth £100, what is £100 in one year's time worth?' Or, putting the question another way, 'what would you accept today instead of the £100 in one year's time?'

The answer depends on the discount rate applied and the period over which it is necessary to discount the future cash flow.

In our example, the present value (PV) of that £100 in a year's time is £90.91. The calculation to arrive at this number is £100/1.1. The 1.1 represents '1 + i', where 'i' is the discount rate expressed as a decimal. So, in our example, 10% = 0.1 and 1 + i therefore equals 1.1.

To confirm that this is correct we can assume that if we have £90.91 now and invest it at 10% the answer would be:

$$£90.91 + £9.09 \text{ (10\% of £90.91)} = £100$$

or:

$$£90.91 + (£90.91 \times i) = £100$$

also expressed as:

$$£90.91(1 + i) = £100$$

so:

$$\text{PV of £100 received in one year} = £100/(1 + i)$$

therefore:

$$\text{PV of each £1 received in one year} = £1/(1 + i)$$

Thus we have a means of converting money received in a year's time into its present value today. £100 received in a year's time is only worth £90.91 today, £1 received in a year's time is only worth £0.9091 today. Using this principle, we can apply a factor of 0.9091 to convert future cash flows to their present value using a discount rate of 10%:

$$0.9091 \text{ is equal to } 1/1.1, \text{ or } 1/(1 + i)$$

Assuming a cost of capital of 10% the discount factors for the first five years will be:

Year 1	0.9091	[$1/1.1$ or $1/(1 + i)$]
Year 2	0.8264	[$1/1.1^2$ or $1/(1 + i)^2$]
Year 3	0.7513	[$1/1.1^3$ or $1/(1 + i)^3$]
Year 4	0.6830	[$1/1.1^4$ or $1/(1 + i)^4$]
Year 5	0.6209	[$1/1.1^5$ or $1/(1 + i)^5$]

DCF TECHNIQUES SUPPORT SHARE VALUATION

Returning to the shareholder decision, we now have a mechanism which enables us to quantify the extent to which the 'money value' capital gain of £50 (purchase price £100, sale price £150) can be converted into a 'real' capital gain. We do this by converting the £150 to be received in five years' time into its present value, to create a true comparison with the £100 being spent today. Using the discount rates above the present value is £150 × 0.6209 = £93.13. This means that, in real terms, there appears to be a loss on the deal.

Let us now add a dividend into the equation. We will assume that a £12 dividend is paid for each year in which the share is owned. Again, the £12 dividend received in a year's time is not worth £12 in present value terms: it is worth £10.91 (£12 × 0.9091). We must therefore make a more complete evaluation of the transaction. We will find that the dividends received will compensate for the real-term loss shown above.

The full discounted cash flow will look like this (note that we are assuming that the sale was made just after the year 5 dividend was received; if not the sale price might have had to be shown in year 6):

Year 1	Dividend	£12 × 0.9091	=	£10.91
Year 2	Dividend	£12 × 0.8264	=	£9.92
Year 3	Dividend	£12 × 0.7513	=	£9.02
Year 4	Dividend	£12 × 0.6830	=	£8.20
Year 5	Dividend	£12 × 0.6209	=	£7.45
Year 5	Share sale	£150 × 0.6209	=	£93.13
Present value			=	£138.63

This can now be compared to the £100 being invested today and the shareholder is shown to be making a positive 'net' present value' (NPV) of £38.63. To the shareholder, a positive NPV means that, if the cost of capital is 10% and if the above assumptions about cash flows prove to be correct, shareholder value will be created.

Thus the concept of present value enables a shareholder looking forward to evaluate whether the purchase is likely to be worthwhile. The DCF model provides a means of valuing the share. In our example the share is worth the present value of the cash flow stream to the shareholder, £138.63, which is more than the price of the share today.

The model is not easy to apply, however. Some companies do not pay dividends, so analysts may substitute the company's earnings or operational cash flows instead. Normally this will be done for a period of three, five or ten years. The sale price of the share in the future is also a difficult item to forecast. Instead analysts may attempt to calculate a single 'terminal value' of the entire earnings stream that will arise after the three-, five- or ten-year period has expired.

In spite of the difficulties with the model it enables share prices to be estimated in a consistent way. The key variables in the equation are the expected future earnings or cash flows of the company. These estimates are produced by analysts on a regular basis and are the lifeblood of professional share valuations.

INTERNAL RATE OF RETURN – THE BREAKEVEN DCF DISCOUNT RATE

In the example above the NPV of the cash flows was £38.63, at a discount rate of 10%. If the discount rate were to be increased to, say, 20% the result would be different because the answer to the key formula '1 + i' would change from 1.1 to 1.2. The NPV of our example would drop to a negative figure, (£3.84). So, using a higher discount rate reduces the present values of the future cash flows. Conversely, using a lower discount rate increases the present value of the future cash flows.

If the share price today were £138.63, as opposed to £100, then the 10% discount rate would generate an NPV of £0 (the PV of the future cash flows, £138.63, less the price of the share today, also £138.63). This is significant. It shows that the 10% discount rate is the one required to break even in DCF terms.

The breakeven DCF percentage is known as the 'internal rate of return' (IRR). It is a useful technique for comparing one series of cash flows, or one share, with another. A share with an IRR of 25% should, all other things being equal, be preferred to a share with an IRR of 15%.

An IRR can be computed by trial and error. This involves solving the formula for two different discount rates and then using interpolation to home in on the real answer. Fortunately, modern spreadsheets have an IRR function and can do this very quickly. In our example the IRR is approximately 19%.

TOTAL SHAREHOLDER RETURN

TSR, or total shareholder return, is the application of IRR to shareholder value. It has been used by companies for many years in the context of investment decisions. It is known by a number of labels – TSR, IRR, DCF rate of return and, perhaps the most useful description of its nature, the DCF breakeven rate.

The meaning of TSR to the investor is explained in Chapters 1 and 2.

RESIDUAL INCOME

In our earlier example, the NPV of the share was £38.63 at a discount rate of 10%. This is equivalent to saying that the share price today, £100, is £38.63 less than it should be. In other words, it is underpriced by the amount of the NPV. Similarly, if the NPV were negative, the share would be overpriced by that amount.

Residual income is a mathematical variation on DCF techniques that enables the under- or over-pricing to be highlighted. It is also the basis for the calculation of economic profit. As such it provides a mathematical link between the valuation of a company's shares and the value it can add by increasing its earnings or cash flow.

Residual income works by deducting a capital charge from the stream of cash flows. The charge is based on the amount of the investment and the cost of capital. In addition the original cost of the share is deducted from the final share price. This gives an identical result to the previous calculation.

The easiest way to illustrate residual income is to include it in our previous example. The capital charge in each year will be £10, which is 10% of the original investment of £100:

Year 1 Dividend	£12 less capital charge £10 = £2 × 0.9091	= £1.82
Year 2 Dividend	£12 less capital charge £10 = £2 × 0.8264	= £1.65
Year 3 Dividend	£12 less capital charge £10 = £2 × 0.7513	= £1.50
Year 4 Dividend	£12 less capital charge £10 = £2 × 0.6830	= £1.37
Year 5 Dividend	£12 less capital charge £10 = £2 × 0.6209	= £1.24
Year 5 Share profit	£150 − £100 = £50 × 0.6209	= £31.05
Net Present Value		= £38.63

Some schools of thought use alternative ways of calculating residual income. These methods base the capital charge on balance sheet assets rather than share price, but they do not produce the identical results shown above.

Calculation of the weighted average cost of capital

This topic is highly complex and, even among experts, opinions differ about the right approach to use. Most agree, however, that the cost of capital has two main components, the cost of equity capital and the cost of debt capital. Both are expressed as a percentage and both reflect different sources of investment funding.

Since most companies are financed by both equity and debt, to arrive at a single percentage it is necessary to compute a weighted average of these two figures, which represents the relative proportions of each type of capital. The resulting figure is known as the 'weighted average cost of capital' (WACC).

Most of the complexity in WACC calculations surrounds the cost of equity finance. Relatively speaking, the cost of debt is easier to calculate.

COST OF DEBT CAPITAL

Strictly speaking the cost of debt is the post-tax rate of return that the holders of debt capital require on their loans. The correct way to compute this cost of debt is to calculate an IRR, in exactly the same way as was discussed in Appendix 1. The cost of debt will be that percentage discount rate which equates the future interest stream and future redemption payment with today's market value of the debt. So the inputs to the IRR calculation are:

■ the current market value of the debt;

■ the interest stream, net of tax, from now until the date when the debt is due to be repaid;

■ the redemption amount due when the debt is repaid.

As an approximation an alternative and simpler method can be used. The cost of debt can be calculated as the average interest rate on the company's loan finance, less tax.

COST OF EQUITY CAPITAL

As with debt capital, the cost of equity capital is the rate of return which satisfies the needs of equity shareholders.

There are many different ways of estimating the cost of equity finance, including the dividend discount model (DDM). This has numerous variants of its own and we have referred to them throughout this book. The model is another IRR calculation, which equates the present value of the future dividend (or EPS or cash flow) stream with the current share price. An example of this model is described in the IRR section of Appendix 1, in which the result is 19%. It is not therefore repeated here.

The Capital Asset Pricing Model (CAPM), which was created in the 1960s, breaks down the return required by equity investors into a risk-free element and a risk premium.

Capital Asset Pricing Model

CAPM is well known because of the 'fame' of one of the particular components of the model. This component is known as 'beta', the Greek letter 'B'. It measures the volatility of the share in relation to movements in the stockmarket as a whole.

CAPM theory states that investors in equities bear a certain amount of risk. Their chosen companies may be loss-making, reducing the value of their investment. They may even fail entirely, going into liquidation and leaving nothing for the shareholders. In addition, share prices are volatile so investment values fluctuate up and down.

To compensate for all the additional risks that equity investors bear, CAPM states that they require an additional return, in excess of that which they could earn by investing in a risk-free asset such as government bonds.

Some of the risk in equities can be diversified away by buying a mixed portfolio of shares. So, on average, when one share goes up another may fall. However, much of the risk is thought to affect all shares in a similar way. So, for example, an economic downturn is likely to depress the profits and share price of most companies. This risk cannot be eliminated by diversification.

The additional return is meant to compensate for this general 'market' undiversifiable risk. CAPM calculates this additional return in two parts. The first is called the 'market premium', which represents the excess return investors have historically required from investing in the whole market as opposed to choosing risk-free investments such as government bonds. The second is 'beta', which measures the amount by which a particular share price is likely to rise when the market rises by a given amount.

So, if the general market return required by equity investors is, say, 13% and the risk-free rate of return in government bonds is, say, 7% then the equity premium is 6% (13% − 7%). For a share with a beta of 1.5, shareholders will require a return calculated as follows.

Risk-free return (government bond yield) =		7%
Risk premium (beta × market premium) =	1.5 × 6% =	9%
Total cost of equity capital =		16%

From the above example it can be seen that there are three components to the CAPM equity cost of capital – the risk-free return, the equity risk premium and beta.

For companies which wish to compute their cost of capital in this way there are many business schools and data providers, such as Thomson Financial Datastream, which publish betas for all quoted shares. In addition the yield on government bonds is widely available. The historic market premium is a matter of frequent academic debate but a figure of 4.5% seems to win the popular vote at present.

The CAPM model is based on long-term research of historical trends of the major stock markets. As such it is prone to error resulting from one-off historical abnormalities that may not recur in the future. At one time the CAPM model was very highly rated but more recent research has thrown doubts on its validity, suggesting a potential tolerance of plus or minus 3%. However, it still remains the mostly widely used measure of a company's cost of equity capital; it is easy to use and the basic concept of linking risk to return remains conceptually sound.

An alternative method, similar to DDM, is called the market derived discount rate. This model approaches the issue from an investor's point of view. It uses market predictions of a company's long-term free cash flow and then equates it, using IRR, to the current value of these same businesses. The argument for this approach is that today's market value reflects the present value of expected future cash flows. Therefore we can work out the return which shareholders require by projecting their future cash flow expectations and relating them to today's market price.

PUTTING IT ALL TOGETHER – WACC

Once the two components of WACC have been calculated it is a simple matter to finish the task. A simple example will show the method.

	Cost of capital (A)	% weight (B)	Overall weight (A × B)
Equity	13%	70	9.1%
Debt	9%	30	2.7%
Total WACC			11.8%

In this case the weighted average comes to exactly 11.8% but it would be unrealistic to believe that any WACC can be accurate to the nearest 0.1% because of the many assumptions built into the model. In practice this spurious level of accuracy creates a false impression of its validity. Certainly the financial directors of major businesses, who often see such numbers published for their companies, regularly dispute the published figures.

The important principle here is to take into account the time value of money. Spurious accuracy should be avoided. There may always be room for dispute about the correct way to estimate WACC but there should never be dispute about the need to use it to help measure and enhance shareholder value.

Common adjustments in economic profit models

This appendix contains a brief description of those adjustments which, while having behavioural implications, are primarily designed to align accounting profit more closely to economic reality. The more common adjustments to arrive at a truer definition of economic profit are as follows:

- *Capitalising research and development and (long-term) marketing expenses and depreciating them over future periods.* The normal accounting convention is to treat such expenditure as an expense in the profit and loss account in the year it is incurred, a method which does not normally reflect economic reality. If a division/company has to absorb an unusual amount of R&D in one year this will reduce profits and hence EP. By spreading the costs and matching them to future income, this adjustment creates a truer definition of profit and also encourages management to maintain investment in R&D for future benefit.

- *Adding acquired goodwill into the capital employed number, where this practice is not followed in normal accounting policies.* If goodwill is included in the definition of capital employed, managers are made accountable for the cost of financing businesses that have been purchased in the past. It could be argued that this is demotivating to managers (particularly if they did not agree the acquisition price) but it is a reflection of economic reality and many companies feel that it should be included in any assessment of shareholder value.

- *Changing the method of depreciation.* In their normal financial accounts the majority of companies use the straight line method based on historical cost, with no further charges once the asset has been fully written off. There are some companies, such as Unilever and Phillips, who use a different method of depreciation for internal management accounting, to reflect the economic use of these assets. This involves indexing historic costs up to current prices and making a notional charge for all assets in use at today's values.

- *Capitalising leases and treating them as if the assets had been purchased and the money borrowed.* Accounting principles already require this adjustment to be carried out for 'financial leases', those which are in reality borrowing transactions by another name. This adjustment extends the principle to all leases, arguing that they are all ways of financing assets used in the business.

Enterprise Value Added (EV+)

EV+ INCORPORATES MANY MODERN VALUATION CONCEPTS

The new breed of DCF valuation approaches, described in Chapter 2, Chapter 3 and Appendix 1, are becoming more widely used by analysts in their reports. It is therefore important that companies who wish to demonstrate superior shareholder value, and in particular the potential for capital growth, understand how these approaches work and the likely impact upon their share valuations of a change in forecast earnings or cash flow.

EV+ is an emerging methodology patented and marketed by Sharevaluer. It builds on two key modern valuation approaches, the DCF residual income valuation model (described in Appendix 1) and the concept of Enterprise Value, or 'EV', which represents the market value of the entire business' equity and debt capital.

THE IMPORTANCE OF ANALYST FORECASTS IN EV+ VALUATIONS

The key variable in all DCF share valuation models is the forecast of future cash flows (or a variant such as dividends or earnings). EV+ uses the earnings or cash flow forecasts produced by analysts as the input for this variable. These forecasts of future performance, available from data providers and elsewhere, are therefore a key driver of share price movements. Changes in specific or consensus expectations should, all other things being equal, have an immediate effect on the value of an equity.

If an analyst alters his forecasts for, say, each of the next three years the only way to measure the combined impact of these changes is to use a DCF technique, such as EV+, which allows for the time value of money.

THE VALUATION OF THE ENTIRE ENTERPRISE IS INCREASINGLY POPULAR

Analysts who use DCF approaches employ forecasts of profit or cash *before interest*. As a result they tend first to value the entire business, including debt. They call this value 'Enterprise Value' or 'EV'. It consists of the market capitalisation of the equity shares plus a market value of the company's debt.

They then have to deduct from EV the market value of debt in order to arrive at their target value for the equity. This is then divided by the number of shares in issue to give a target share price.

Analysts prefer to value EV, as opposed to market capitalisation, because they can use forecasts of earnings or cash as the input to their models. They do not need to make adjustments for interest payments first. It makes intuitive sense to produce a value for the entire business, irrespective of its capital structure, before attempting a valuation of its shares.

EV is popular for another reason too. It lends itself to the production of comparative performance ratios such as EV/EBITDA and EV/sales. These ratios are becoming more widely used since they measure the relationship between performance and market value. They are gradually replacing ratios based upon the balance sheet value of capital employed, which are fraught with the difficulties of historic accounting valuations.

EV+ IS CLOSELY LINKED TO ECONOMIC PROFIT

Like Economic Profit, EV+ is a residual income technique, so the model can help those companies who wish to translate the market's expectations of share price performance into internal value-based targets.

Table A4.1 below shows how a business, Company A, can be valued using EV+ methods. The process is similar to the residual income example in Appendix 1. This particular example uses an analyst's free cash flow forecasts for three years and the analyst's terminal value of the company, £0.9m, representing the value of the free cash flows from Year 4 onwards. The current market value of the company's equity is £0.5m and the debt is £0.2m, so the EV is £0.7 (£0.5m plus £0.2m). The company's cost of capital has been calculated to be 11 per cent. It has 100,000 shares in issue so the current price per share is £5.

In Table A4.1 it can be seen that the value of the firm will increase as the forecasts of free cash flow increase.

It should be obvious that this valuation model can also be used by companies when they set strategic financial targets for earnings and cash flow. It enables them to model the impact of different scenarios.

However this method of valuation is much more than a simple DCF-based technique. It also aids comparative valuations with an entire index, a sector or a peer group of similar companies.

Table A4.1 Company A valuation using EV+

	Yr 1	Yr 2	Yr 3	PV Total
Free cash flow (£000)	140	150	165	
Less capital charge at 11% of EV, £0.7m	(77)	(77)	(77)	
Residual income (RI) stream	63	73	87	
Discount factors @ 11%	0.901	0.812	0.731	
EV+ stream (RI × discount factor)	56.8	59.3	63.6	179.7
Add terminal value profit (£0.9m – £0.7m)			200	
Discount factors @ 11%			0.731	
PV (terminal value profit × discount factor)				146.2
Total under/(over) pricing of company (£000)				325.9
Add current Enterprise Value (£000)				700.0
Estimated true Enterprise Value (£000)				1025.9
Less market value of debt (£000)				200.0
Estimated true value of Equity Capital (£000)				825.9
Number of issued shares				100,000
Target price per share (Value of equity/issued shares)				£8.26

EV+ ASSISTS RELATIVE PEER GROUP VALUATIONS

Using the above cash flows for Company A, the first step in a relative valuation is to calculate the PV of the residual income stream (EV+) as a ratio to share price. This generates a ratio that is directly comparable with other companies. Table A4.2 shows the ratios for Company A compared to a similar set of ratios for an appropriate comparator peer group (the calculations for the peer group are not shown but they are identical to methods used for Company A).

Table A4.2 Company A valuation using EV+

	Yr 1	Yr 2	Yr 3	PV Total
Free cash flow (£000)	140	150	165	
Less capital charge at 11% of EV, £0.7m	(77)	(77)	(77)	
Residual income (RI) stream	63	73	87	
Discount factors @ 11%	0.901	0.812	0.731	
EV+ stream (RI × discount factor)	56.8	59.3	63.6	179.7
EV+ per share (EV+/100,000)	£0.57	£0.59	£0.64	
Price/EV+ (£5:EV+ per share)	8.8	8.5	7.8	
Peer group mean P/EV+	7.5	7.2	7.6	
% difference (Co. A / Peer group)	17.3	18.1	2.6	

The initial valuation of Company A showed that it appeared undervalued. Its current price of £5 is less than its target price of £8.26. The total undervaluation consists of the EV+ stream and the PV of the terminal value profit. It results in positive values for the P/EV+ ratio. If the EV+ stream were negative the share would be underpriced and the P/EV+ ratio would also be negative.

A similar analysis of a peer group of companies shows that they too are undervalued. Therefore it is possible that the entire sector has been downrated, perhaps for general economic reasons, and this needs to be reflected in the valuation of Company A. The way to incorporate this downrating is to measure the relative ratios of P/EV+. The comparisons in Table A4.2 show that Company A's ratios are between 2.6 per cent and 18.1 per cent higher than the peer group. The mean of these is 12.6 per cent. All other things being equal this indicates that, compared to its peers, Company A is undervalued by around this figure. So its true share price in the current market should be around £5.63. This is a relatively short-term valuation as it focuses on EV+ for three years only. As the market for the sector improves this target price may rise towards the initial valuation of £8.26.

There is yet another factor which could be considered before finalising the target price. The valuation above only focuses on the short-term relative EV+ streams of Company A and its comparator peer group. It ignores the relative terminal value profits, which are longer term in nature. It is possible that the terminal value profits of the peer group, when compared to share price, may be significantly different from Company A. If so, this difference should also be taken into account.

By combining the ratios to price of both the EV+ stream and the terminal value profits, it is possible to identify the overall relative over/underpricing of a company in relation to its peers. Similar calculations can be performed for the comparison of entire sectors.

EXTENDING THE USE OF EV+

EV+ lends itself to the production of other comparative metrics in addition to the ratio P/EV+. For example, just as with PEG the growth rate in EPS is a vital variable, it is possible to produce a similar ratio for EV+, incorporating the growth rate in EV+ from one year to another. This also helps to identify those companies that are significantly mispriced compared to their peers and highlights those that will become increasingly so if they achieve their forecast earnings growth rates.

The calculation and use of the EV+ PEG (P/EV+G) can be illustrated using the same numbers as before (*see* Table A4.3).

Table A4.3 Company A calculation of P/EV+G

	Yr 1	Yr 2	Yr 3	Mean
EV+ stream	56.8	59.3	63.6	N/A
Growth rate	N/A	4.4%	7.3%	5.8%
P/EV+	8.8	8.5	7.8	8.2
P/EV+G (P/EV+ divided by growth)	N/A	1.9	1.1	1.4

This ratio is also useful for comparative share valuations. As with normal PEG ratios, the higher the growth rate the lower the ratio. So, faced with the choice between two companies that have similar positive P/EV+ ratios, investors should prefer the one that has a lower ratio of P/EV+G.

In the above calculations the mean figures are calculated as follows. The growth figures are a compound interest figure. The P/EV+ mean only incorporates years 2 and 3.

There are many other metrics that can be developed from these principles. For example, another way of incorporating growth in EV+ is to calculate the percentage growth that is required by a particular company in order to justify its value in relation to its peers. Alternatively it is possible to calculate a break-even point, measuring the number of years it takes for a company to generate a positive value in DCF terms. This can again be compared with peer groups of similar companies to identify under- and over-pricings.

In addition to appreciating the variety of metrics EV+ enables, it is also important to recognise that analysts can use these techniques for purposes other than straightforward stock valuation. The obvious extension of the model is to use the method to measure the relative value of different sectors, which aids the asset allocation decision faced by fund managers. However, it also lends itself to sensitivity analysis of valuation conclusions and to the measurement of the statistical significance of mispricings.

Finally, because EV+ incorporates forecasts of cash flows or earnings, it is possible for analysts to substitute their own in-house forecasts rather than published consensus estimates. This can generate unique insights that other analysts may not have.

THE CONCLUSIONS FOR COMPANIES

The arrival of sophisticated techniques like EV+ presents a challenge for companies aiming for shareholder value. Understandably they will seek simplicity in the targets they adopt internally. However, they will also need to stay ahead of the analysts in predicting the implications for share price of changes in their forecast performance.

They will need to build models that show the sensitivity of their market capitalisation. This means that they will need to generate some expertise in techniques such as EV+, using it to compare their own value with their peers on a consistent and regular basis.

Examples of analyst share valuations

This appendix discusses examples of two valuation reports. The first, by Paribas, demonstrates a valuation relative to a peer group. The second, by ING Barings, adopts an absolute DCF-based approach.

Fig. A5.1 Paribas relative share valuation, 18 February 2000

ñ PARIBAS

| 18 February 2000 | Engineering – UK |

Buy
European Research Alert

Key data

Price	272p
52W Hi/Lo	194p/358p
Market cap.	Euro16,309m
Major shareholders	Franklin 6.0%
	Capital 5.0%
Free float	85%
Derivatives	none
Index (FTSE 100)	6147
Reuters code	ISYS.L
Bloomberg code	ISYS LN

Financials *

	99	00E	01E	02E
PBT (£m)	295	648	993	1287
Rec EPS (p)	17.8	20.4	22.6	25.8
Rec CFPS (p)	20.7	11.7	18.9	32.6
DPS (p)	6.2	7.5	8.3	9.1
Rec PER (x)	15.3	13.3	12.0	10.5
Recurrent PCF	13.1	23.1	14.4	8.3
Div yield (%)	2.3	2.8	3.0	3.3
Price/Book (x)	1.5	1.5	1.3	1.2
EV/EBITDA	8.5	7.8	7.0	6.7

* Year ended 31 March

Earnings Dynamics

Price (£) and price relative

Author

Name: Simon Fenwick
+44 171 595 3235
Location: London
simon_fenwick@paribas.com

Invensys Plc

EPS growth 12%pa to 2002. Longer-term upside from home automation growth.

In the short term we expect the Invensys share price to stabilise at current levels as expectations for an acquisition increase. However, in the longer term, share price appreciation should be underpinned by (1) our forecast for 12%pa EPS growth over the next 3 years, driven by the fruits of restructuring and efficiency programmes, and (2) growth within the automation division, as it participates in the potentially huge global market for home automation systems. Invensys shares are trading at a 15% discount to the peer group and a 63% discount to the UK market based upon our forecast 2002 PER. Our price target is 360p.

Key points

- Invensys is nearing the end of a restructuring programme that has transformed the group from an engineering conglomerate into a focussed provider of automation, control, power and industrial drive systems.
- Recent share price weakness can be linked to unconfirmed rumours of a bid for Schneider. An acquisition of Schneider, although complementary, would strengthen Invensys's hand, via an enlarged European market share, should Invensys eventually merge with a US competitor.
- Invensys is trading on attractive multiples without factoring in the longer-term upside from the automation division, which is set to participate in the last wave of the internet revolution – automated home environments linked with service providers by internet. We estimate the size of this market at $20bn pa by 2010.
- Our price target for Invensys is 360p/share, representing upside of 32% from the current share price. We rate the shares as our preferred long-term holding within the Paribas 'engineering' sector.

Summary

Focussed – Controls &automation now accounts for 75% of Invensys's sales, making it, along with Emerson (75%) and Schneider (78%), the most focussed companies within its global peer group. Cost savings from the BTR/Siebe merger and the 'Six sigma' efficiency programme should combine to boost PBT by £300m pa by 2002.

Acquisition trail – The £1.8bn divestment programme is now 80% complete and Invensys will have a £3.0bn war chest to spend over the coming years. An acquisition of Schneider (market cap £7.0bn) would provide Invensys with shared global leadership in industrial controls with Siemens, but many of the businesses are complementary in their profiles and merger synergies would not be high. Alternatively, Williams of the US or the security division of troubled conglomerate Tyco could be targets to fill the gap for Invensys on security and fire-protection systems. Should Invensys acquire one or more of these targets it would be well placed should it eventually merge with fellow market leader Emerson. This deal would deliver significant synergies.

The Automation market – Invensys is the global #1 in residential automation. Whilst this market is not likely to take off for 5 years, its potential size is significant. The potential revolves around controlling the flow of all incoming services and utilities into a home via integrating automated controls with the internet. Based upon a retail price of roughly $1,000/unit for Invensys's automation software provided to these systems, the market could grow to $20bn pa, assuming a conservative 20% penetration of automation systems into US households over the next ten years.

Valuation – Invensys is trading at a 17% discount to the global automation and control sector* based upon our 12.0x PER estimate for 2001 and a 15% discount based upon our 10.5x PER estimate for 2002. **We have used the unweighted average multiple for Emerson, Rockwell, Honeywell and Schneider and adjusted for the financial year-end.*

BNP PARIBAS GROUP

Figure A5.1 shows a relative share valuation by Paribas. It is, of course, only an extract of the full valuation. However, even the extract shows the use of a wide variety of data, including:

- textual analysis about Invensys and the markets within which it operates;
- facts about the share price, such as market capitalisation and the 52-week high/low;
- three-year forecasts of items such as profit before tax (PBT), dividend yield, EV/EBITDA and P/E (or 'PER');
- a price chart relative to the market.

It is interesting to note the way in which this valuation has been prepared. The key features are as follows:

- Paribas have computed forecast P/E ratios (PER) for the years 2001 and 2002 for both the UK sector and an international peer group of comparable companies (Emerson, Rockwell, Honeywell and Schneider). These PERs are simply a comparison of the current price with the forecast EPS for the years concerned.
- Paribas have then compared these aggregate PER forecasts with those of Invensys for the same years to establish that Invensys is trading at a discount. The price at that date was 272p and the FTSE 100 was trading at 6147.
- They have estimated a target value for the share (360p) assuming that some of the discount represents an underpricing which should rectify itself in the market.

This valuation is typical of many in the way that it uses both text and a variety of data items. Its key feature, for us, is the use of the peer group for comparative purposes.

Next we consider the report shown in Figure A5.2 which calculates an absolute valuation, without the use of a peer group. Most valuations of this sort use one or other of the family of DCF techniques. This particular report employs an EVA approach (Stern Stewart's version of economic profit), which is one of those variants.

In this case the analyst is valuing the company's market value added (MVA). The formula for MVA is as follows:

Market value of equity and debt less Balance Sheet value of equity and debt

MVA represents the amount that the management have added to what has been invested and is therefore a reasonably good measure of value creation. What makes MVA even more significant is its close relationship to economic profit. Some argue that:

Current MVA = Present value of all future economic profit

Fig. A5.2 ING Barings absolute valuation, 26 January 2000

Perhaps the next deal that Invensys should look to make is a full merger with Emerson for maximum shareholder value creation, and this would explain nicely why Invensys' current management team continues to grumble about its discount rating relative to its major peers!

EVA VALUATION

Figure 2			EVA VALUATION (£m)					
Yr to Dec	1999	2000F	2001F	2002F	2003F	2004F	2005F	Cont value
Shareholder funds	7,581	7,867	8,384	9,003	9,633	10,308	11,029	11,801
Minorities	236	180	189	198	208	219	230	241
Provisions	788	764	741	719	698	677	656	637
Cumulative goodwill written off	0	0	0	0	0	0	0	0
Real equity	8,605	8,811	9,314	9,921	10,539	11,203	11,915	12,679
Convertible equity	0	0	0	0	0	0	0	0
Net debt/(cash) inc convert as debt	1,976	1,283	1,083	778	578	378	178	0
Total capital	10,581	10,094	10,397	10,699	11,117	11,581	12,093	12,679
Operating profit	1,008	1,217	1,260	1,365	1,474	1,592	1,720	1,857
Tax charge	(390)	(315)	(355)	(390)	(422)	(455)	(492)	(531)
Less: interest tax shield	(67)	(45)	(31)	(20)	(10)	(2)	0	2
NOPAT	551	857	874	956	1,043	1,135	1,228	1,328
Return on capital (%)	5.2	8.5	8.4	8.9	9.4	9.8	10.2	10.5
Cost of capital (WACC = 9.0%)	9.0	9.0	9.0	9.0	9.0	9.0	9.0	9.0
EVA	(401)	(52)	(61)	(7)	42	92	139	187
Discount factor	1	0.917	0.842	0.772	0.708	0.65	0.596	0.547
Discounted EVA	(401)	(48)	(52)	(6)	30	60	83	102

Opening capital	10,581	
EVA forecast period	68	**Continuing value assumption**
EVA continuing value	4,911	2% spread, 6% growth rate
Net debt & minorities	(1,103)	
Equity value	14,457	
Equity value per share (p)	**389**	

Source: ING Barings estimates

From the above EVA calculation we believe that Invensys could be worth up to 390p per share, some 22% above the current share price of 320p. This value is arrived at by assuming a return on incremental capital of 2% above the ongoing WACC of 9%. In addition, we assume a growth rate of 6%, a little ahead of the companies' expectations of some 5% growth from the combined BTR and Siebe businesses. Note: if we were to assume a 5% growth rate then the implied EVA valuation falls to 367p.

After valuing MVA the analyst then deducts the value of debt to arrive at an 'equity value' or market capitalisation. This is actually also an enterprise value-based approach, looking at the implied *market value* of the company. Once the analyst has calculated equity value it is then a simple matter to calculate a target price per share, dividing the equity value by the number of shares in issue.

The valuation uses forecasts for the period from 2000 to 2005. It then adds a 'continuing value' to represent the years from 2006 to infinity.

The key steps in this valuation are the calculation of:

- total capital – the sum of shareholders' funds, debt and other items;

- NOPAT – net operating profit after tax;

- EVA – NOPAT less the charge for the cost of capital (total capital × 9%);

- discounted EVA – by discounting and adding the individual EVAs the analyst produces an EVA sum (MVA) for the forecast period (68) and for the continuing value (4911);

- equity value – this represents the opening capital, plus MVA, less debt.

This sort of approach is very different from the first. It is fairly typical of the DCF-type calculations found in analysts' reports. The method produces an absolute valuation and does not rely on peer groups.

Both of these valuations are, of course, significantly out of date yet they are good illustrations of the main approaches analysts take. Most reports use a mixture of both techniques so companies which wish to demonstrate shareholder value would be wise to understand them.